XVII Biennial Congress of the Argentinean Bioengineering Society and VI Clinical Engineering Conference 2009

(SABI 2009)

Journal of Physics: Conference Series Volume 313

Rosario, Argentina
14 – 16 October 2009

ISBN: 978-1-61839-132-2
ISSN: 1742-6588

Printed from e-media with permission by:

Curran Associates, Inc.
57 Morehouse Lane
Red Hook, NY 12571

Some format issues inherent in the e-media version may also appear in this print version.

Printed by Curran Associates, Inc. (2011)

For permission requests, please contact the Institute of Physics
at the address below.

Institute of Physics
Dirac House, Temple Back
Bristol BS1 6BE UK

Phone: 44 1 17 929 7481
Fax: 44 1 17 920 0979

techtracking@iop.org

Additional copies of this publication are available from:

Curran Associates, Inc.
57 Morehouse Lane
Red Hook, NY 12571 USA
Phone: 845-758-0400
Fax: 845-758-2634
Email: curran@proceedings.com
Web: www.proceedings.com

TABLE OF CONTENTS

PAPERS

PREFACE

SABI 2009 was the XVII Biennial Congress of the Argentinean Bioengineering Society (SABI – www.sabi.org.ar), celebrated along with the VI Clinical Engineering Conference. It took place in Rosario, the second city of Argentina, located on the west bank of the Paraná, one of the world's most important rivers. This city, with its 150 year history and one million inhabitants, is characterized by a strong enterprising spirit. It is the agroindustrial leader of Argentina, with cereal ports recognized to be among the most active in the world, and its cereal stock exchange competes with Chicago's in international cereal pricing. Demographically Rosario presents a European profile, and there are seven national and private higher level universities in the area.

SABI 2009 was the first time the Congress was celebrated in Rosario. Usually the Congress is organized by the Bioengineering Society in cooperation with a university with an undergraduate program, which Rosario lacks. To meet the needs of this exceptional case, a young local institution was asked to coordinate the Congress, the Rosario Technological Center (www.polotecnologico.net). This organization gathers together around 100 companies that produce technology, with a large number focused on IT, but those focused on biotechnology also stand out. The Center is also integrated with relevant public and government bodies.

Traditionally, bioengineering has been related to human health applications, with less emphasis on applications significant to agrotechnology, an area in which Rosario is growing as an economic force. In order to address this oversight, the Congress formulated its main goals for integrating and synergizing bioengineering and biotechnology, particularly bioengineering and agrotechnology. This initiative has produced promising results.

The importance of the Congress was reflected in the high number of participants – including researchers, professionals and students – from abroad, with participants from at least 12 different countries. In the three days of planned activities, more than 800 participants were registered and over 2000 people visited the associated expo. Several symposia and thematic meetings were held simultaneously.

Around 207 works were presented, each evaluated by two referees, and 181 presentations were finally accepted. The organizers of the Congress proposed that the 14 best works be published in the *Journal of Physics Conference Series*.

A few of the speakers deserve a special mention: James Lin, PhD, USA, *Wireless noninvasive sensing and physiological signatures and vital signs*; Luis Kun, PhD, USA, *Biomedical Engineering and Global Health Transformation: A vision of opportunities and challenges*; Mario Munich, PhD, USA, *Avances en Robótica y Aplicaciones en Salud* (Advances in robotics and health applications); Jose Folgueras, PhD, Cuba, *Generación de productos de alto valor agregado a través del conocimiento* (Generating products adding value through knowledge); Daniel Goroso, PhD, Brazil, *Avanço e tecnología en Medicina Física y Rehabilitación* (Advances in technology in physical medicine and rehab); Adolfo Spangenberg, PhD, Brazil, *Telemedicine and Health – New perspectivas in development countries*; Francisco Rojo Pérez, PhD, Spain, *Propiedades del Comportamiento Mecánico y la rotura de la aorta humana* (Properties of mechanical behaviour and rupture of the human aorta); and Eduardo de Forteza, PhD, Argentina, *La fusión del humanismo médico y la tecnología* (Fusion between medical humanism and technology), among others.

We would like to thank the institutions, universities, government bodies, and private enterprises that supported and facilitated this Congress, and the members of the Organizing Commitee: Eng David Astegiano (Science and Technology State Minister), Eng Carlos Cerruti (Past President of the Rosario Tecnology Center), Eng Viviana Rotger (Past President of SABI) and MSc Guillermo Civetta (Manager of RTP).

The members of the executive committee also deserve special recognition. They have made up for their lack of experience with their great effort and dedication, achieving a level of excellence: Melina Rua, Cecilia Rocha, Noelia Vergonzi, Guillermo Civetta, Ricardo Montico, Ariel Olmos, Guillermo Chavez, Marcos Vaccani, and others.

Finally we thank all the presenters, speakers, exhibitors, partners and everyone who contributed to this successful Congress.

The next gathering will be held in the internationally known city of Mar del Plata on 28–30 September 2011. It will undoubtedly be a continuation of this notable event which has been held annually for 30 years.

Ing Darío Rocha
SABI and Polo Tecnologico Rosario Committees Member

Journal of Physics: Conference Series **313** (2011) 012001 doi:10.1088/1742-6596/313/1/012001

Early detection of ventricular tachycardia with sending messages to cell phone

L J Ramirez, F A Lozano and C R Rondon
Telemedicine Research Group of the Nueva Granada Military University. Bogotá, Colombia.

tigum@unimilitar.edu.co

Abstract. Sustained ventricular tachycardia (VTs) can be asymptomatic for some people, but for other is deadly because it is a major cause of sudden cardiac death [1]. Some patients may present this arrhythmia, and even so, they decide to drive car increasing the likelihood of VTs, putting at risk not only his life but that of the other drivers. We developed a system for early detection of VTs, consisting of EKG sensors, a card of processing and a cell phone, which detects this arrhythmia, gives an alarm signal to the driver, and it simultaneously sending to text messages a specialist doctor and a relative or friend, all in real time. This design was conditioned to the car, is light and comfortable, that allowed that work of car's driver without discomfort. This system will save lives, since in case of emergency sends a help message, no matter where you are in the driver.

1. Introduction

According to statistics conducted by the American Heart Association (AHA) [2] in the world 27.7% of deaths are caused by heart disease, particularly in Colombia is estimated that deaths from this cause are 30% where 8% of these are due to sudden death caused by cardiac abnormalities. Moreover, the elderly are more prone to arrhythmias such as ventricular tachycardia, especially when they are doing activities that generate high stress such as driving, due to heavy traffic and the recklessness of other drivers. Whereas this situation has caused in the past three years more than 10% of car accidents, one in Bogota and 4% of these accidents, accidents causing large-scale [3].

From a clinical standpoint tachycardia may have a ventricular or supraventricular origin and the differentiation is extremely significant, since this depends on the treatment to try to reverse the arrhythmia in the acute phase, the long-term management and prognosis of the patient same [4-6]. Consequently the causes involved in sudden cardiac death are diverse. In that case have been implicated supraventricular tachyarrhythmias, bradyarrhythmias, hemodynamic ischemia and obstruction, but are the predominant cause of malignant ventricular arrhythmias [7-12], that is why in patients considered high risk are monitored constantly. Regarding to development of devices for automatic analysis of cardiac abnormalities has increased in recent years. Currently, the devices automatic of detection ADD are equipment of microprocessor-based devices, highly complex, and then break that record the EKG signal to determine whether it is compatible with ventricular fibrillation (VF) or ventricular tachycardia (VT) without pulse [13].

In fact, with this framework is need to developing an electronic device that can capture the subject's electrocardiogram signals, processed to detect when it happens arrhythmic heart rate and once classified as ventricular tachycardia, resulting in an alarm issue the subject and simultaneously send alert messages to two, the medical specialist and a family. This is how the development of this innovative device is beneficial for all those who suffer cardiac arrhythmias such as ventricular tachycardia is because it gives confidence to enable them to perform their daily activities, such as driving a car, unaccompanied by any person. When an accident happened, this device facilitated the research process, because it carries a record of cardiac abnormalities through text messages sent.

2. Materials and Methods

First, based on current requirement, initially prepares a list of design requirements to make of device functional.

2.1. Materials

Adaptive power supply to a car, portable, lightweight, detecting 140 bpm, alarm only sent by the user is authorized, using a massive communication system to send to alarm.

Then, the tests are performed with ten subjects elderly that have been, although only once, diagnosed with arrhythmia, and they still are drivers of cars. The type of EKG recording is noninvasive, consisting of three surface electrodes located in the chest, these subjects recorded P-R interval between 0.19s to 0.12s, and between 70 to 80 bpm in steady state. In fact, the pre-processing of data, starts with an analysis of uncorrelated noise, thus implementing a band pass filter fs=5Hz, fi=0.33Hz and fc=2.665Hz, in addition of elevator level of twice the original signal amplitude. Therefore, the data processing is done with PIC 16F877A Microcontroller with USART communication port and Bluetooth transmission to module Parani ESD 200 that meets 802.15 protocols. Then, into cellular phone 3G programming, we implement to API jsr 82 for receiving, and API jsr 120 for sending messages.

2.2. Method

This work has developed in stages, as shown in Figure 1. The first stage involves the acquisition of the subject's electrocardiogram recording, electrode placement, signal transmission to the electronic card; the second stage, performed the pre-processing of signal: filtering, amplification and dc level change; the third stage, the signal processing performed: sampling, detection of heart rate and thresholding of information; the fourth stage, is performed by Bluetooth communication protocol 802.15 to a cell phone ; the fifth stage, takes place programming to receive and store information in cell phone and send alarm messages to selected phone numbers, and in the sixth stage is verified that the information of the date and time of delivery was transmitted and stored.

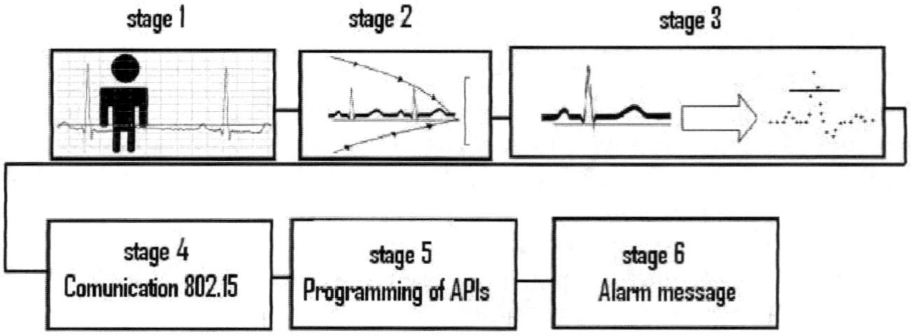

Figure 1: Development stages of the detection system and send alarm

2.2.1. Acquisition of EKG

The acquisition circuit design employs three leads I, II and III, and a gain resistor RG regulating the gain of the system by equation (1):

$$G = 1 + \frac{50k\Omega}{R_G} \qquad (1)$$

2.2.2. Pre-processing data

In the EKG signal preprocessing were implemented several techniques to refine the signal to analyze.

- filtering
  As a result of noise due to the high frequencies and interference from the mains, 60Hz, we design a band-pass filter which limits the frequency spectrum of the EKG between 0.5 Hz and 150 Hz. In addition, we implemented a notch filter that removes narrowband interference from 60Hz. This conditioning also helps to avoid the phenomenon of overlap in the intervals of scanning. Finally, another low-pass filter that eliminates the noise from the signal conditioning.
- Voltage level-shifting
  The output from the preamp brings EKG signal referring to the common floating in order to make the link between references to ground, it is essential to raise the reference voltage signal. Given that, the EKG signal has variations in the range of + / - 1V, we developed a voltage level-shifting of 2V for that change this reference make to EKG signal can be vary within positive values.
- Processing data
  Now, for QRS detection using the A / D converter implemented in the 16F873A microcontroller with 4 MHz crystal. Thus, each instruction takes the time to 4 periods, i.e.:
  $F_{crystal}$ = 4 MHz
  $T_{crystal}$ = 1/4Mhz= 0.25 μs.
  $T_{instruction}$ = 4 x 0.25 μs = 1μs. (2)

  Given that ventricular tachycardia cardiac pulsation is fast, this is characterized by 3 or more consecutive premature ventricular beats. Thus, the analysis is to take the time it takes the development of 3, 5, 10 and 20 QRS complexes so we can determine the heart rate of each subject, as shown in Table I.

Table 1 Registration time at different bpm

Time (s)	**10 bpm**	**3 bpm**	**5 bpm**	**20 bpm**
1	600	180	300	1200
2	300	90	150	600
3	200	60	100	400
4	150	45	75	300
5	120	36	60	240
6	100	30	50	200
7	85,714	25,714	42,857	171,428
8	75	22,5	37,5	150
9	66,667	20	33,333	133,333
10	60	18	30	120
15	40	12	20	80
20	30	9	15	60
25	24	7,2	12	48
30	20	6	10	40
60	10	3	5	20

Now, in Figure 2 shows the barrier of 140 bpm, located exactly in the 4.28s. This means that the time of 10 QRS complexes takes less than 4.28s. This meaning the possibility that the subject is suffering VT. Therefore, with this data we made the decision about how many pulses take to detect VT and also to determine the total time of detection, by the equation:
t_{10}= 60 bpm x 10 p/140 bpm = 4.28 s (3)

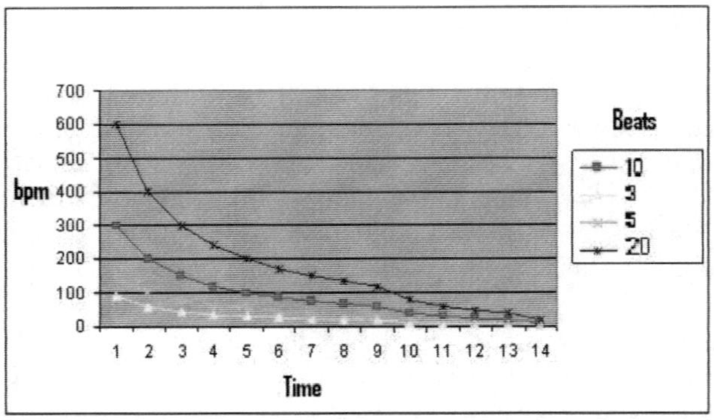

Figure 2: Registration of duration to different beats

- Communication 802.15
  Subsequently, to establishing the connection between the electronic card processing and cell phone we was used Parani ESD module because it allows the port USART interface between the microcontroller and Bluetooth protocol 802.15. This connection has the following characteristics: baud rate 9600bps, no flow control, no parity and no authentication. In addition, to establish Bluetooth connections from mobile devices, we program in J2ME Java language, using the settings for devices with limited connection (CLDC) that maintains, if possible, the characteristics of Java on devices limited to 16 processors or 32-bit , aged 8 to 32 MHz and a total amount of memory available for Java platform at least 160-193 Kb.
- Programming
  The Application Programming Interface API, used in programming the cellular telephone was JSR-82, which is divided into two parts: *javax.bluetooth* and *javax.obex*. Both packages are fully independent. The first defines basic classes and interfaces for device discovery, service discovery, connection and communication. The second protocol can handle high-level OBEX (Object Exchange) used mainly for file sharing, where cell phones are seen as the client in the connection.
- Information register
  At last, in this stage was developed in the language J2ME, with API JSR-120 for exchanging messages over wireless networks. Thus, from this API you program the SMS short message content and determine which users will alarm. Furthermore, keep a record of the date and time the messages were sent and which received a recorded number of received responses.

3. Results

The results show that the developed system is functional and can be installed in any car because it meets the requirements requested, that is to say, adaptive power supply to a car, portable, lightweight, detecting 140 bpm, alarm only sent by the user is authorized, using a massive communication system to send to alarm.

Tests confirmed to us the different values given to the programming of devices for early detection of VT. Thus, with regard to threshold the Figure 3 shows one of this tests undertaken to confirm the optimal cut-off level programmed into the microcontroller include allowing a greater degree of reliability when QRS complex detection. Nevertheless this threshold can be changed for each subject according to their corporal composition and strength of its beats, even so, for subjects studied at early-detection system of VT operated optimally.

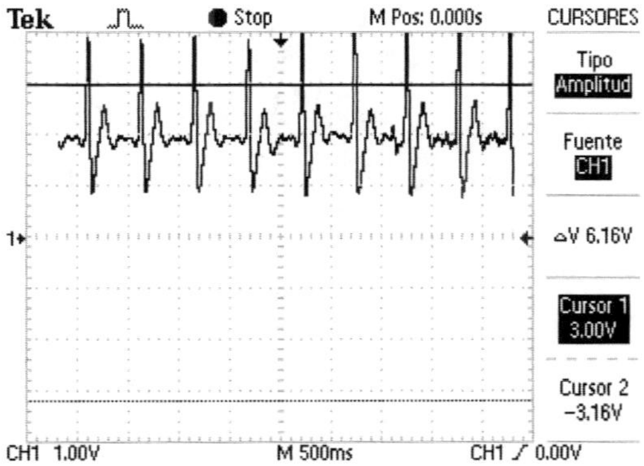

Figure 3: Cut-off level for the detection of the QRS complex

Furthermore, we tested for reliability in the early detection of VT now on the minimum number of beats necessary to establish a limit. Table II shows the results when comparing EKG simulator with a real subject. As a result of testing the system works correctly in the early detection from 140 bpm.

Table 2 Comparison tests between the simulator and a subject

User	**Signal**	**bpm**	**Alarm**	**Sent to SMS**
Simulator EKG	EKG normal	40	NO	NO
Simulator EKG	EKG normal	60	NO	NO
Simulator EKG	EKG normal	80	NO	NO
Simulator EKG	EKG normal	100	NO	NO
Simulator EKG	EKG normal	120	NO	NO
Simulator EKG	EKG normal	140	SI	SI
Simulator EKG	EKG normal	180	SI	SI
Subject	Normal	70	NO	NO
Subject	agitated	140	SI	SI

Regarding the time spent by the system to determine whether the subject has VT. The system analyzes ten beats in a row, then measures the time of these ten pulses and thus the present data analysis for 95% certainty of presence VT. Table III shows that the detection is more delayed when the subjects undergoes abrupt changes your heart rate because the system takes more time analyzing this sudden change in the signal. Otherwise, when the subject has a heart rate progressively increases the system detects the anomaly and immediately sends the alarm.

Table 3 TIME OF DETECTION Simulator VS Subject

User	**Initial heart rate (bpm)**	**Final heart rate (bpm)**	**Alarm**	**Time of detection (s)**
Simulator EKG	60	60	NO	-
Simulator EKG	60	100	NO	-
Simulator EKG	60	140	SI	4.5
Simulator EKG	80	140	SI	4
Simulator EKG	100	140	SI	3
Simulator EKG	120	140	SI	1
Simulator EKG	60	180	SI	4.5
Simulator EKG	140	180	SI	1
Subject	120	140	SI	1
Subject	140	140	SI	1

Finally, with respect to the generation of emergency message could not be established that is the exact time of the generation of the message, because this process takes place in milliseconds and we did not have a method that allows us to synchronize the device with cell phone to compare these times. However, in terms of time taken to send the message, was measured with a stopwatch from the time the alarm is activated when you listen to or view a new alarm message in the cell phone of the person will help to subject. As shown in Table IV.

Table 4 TIME spent in sending a message

Number of messages	**Mobile operator output**	**Mobile operator destination**	**Time (s) spent in sending**	**Average (s) time spent in sending**
1	Operator 1	Operator 1	9.2	9.2
2	Operator 1	Operator 1	9	9,1
3	Operator 1	Operator 1	10.3	9,5
4	Operator 1	Operator 1	9.1	9,4
5	Operator 1	Operator 1	8.6	9,24
6	Operator 2	Operator 1	15.8	10.3
7	Operator 2	Operator 1	20.5	11.8
8	Operator 1	Operator 2	19.3	12.7
9	Operator 1	Operator 2	14.6	12.9
10	Operator 2	Operator 2	8.4	12.5

When messages are sent between the same operators generated text messages arrive in an average of 9s to their destination, whereas when the messages generated are sent to different operators take an average of 18s to arrive successfully at the person who will help test's subject.

4. Discussion

Once the system tests for early detection of VT can confirm that when a heart rate above 140 pulses / min [14-16] without having made any physical effort the patient is at risk of suffering ventricular tachycardia, which was the appropriate criterion for selecting the threshold of detection to provide early warning.

In spite of the early system functions optimally, still the problem of knowing where it is geographically the subject who reported the alarm. Indeed, a solution can be reported by Dr. Haroon Rashid [17] which suggests the use of modulo Global Positioning System (GPS) some 3G equipment, so that when generating the emergency message that contains the user's exact locations.

With respect to the power supply, we have had some minor problems with the energy that is taken from the car, this due to the noise factor. We suggest that to avoid these inconveniences, it is necessary to use a portable power supply (external battery) the system of early detection of VT.

Analyzing Table 1 we could determine that when taking a smaller number of beats for the detection of VT, this detection will take less time. Nonetheless, it increases the number of false positives the system, causing a glut in the output of alarm messages. For this reason, we suggest detect VT with 10 pulses as the time despite being a bit more for the less pulses the error is much smaller.

Acknowledgments
The authors of this paper wish to acknowledge the support given by the Vice-Rector of Research of the Nueva Granada Military University and the Dr. Norman Jaimes for their support in testing.

References

[1]Olgin JE, Zipes DP. 2007. Specific Arrhythmias: Diagnosis and Treatment. In: Libby P, Bonow RO, Mann DL, Zipes DP. *Libby: Braunwald's Heart Disease: A Textbook of Cardiovascular Medicine*. 8th ed. Philadelphia, Pa: Saunders Elsevier; chapter 35.

[2]AHA. 2007. Heart Disease and Stroke Statistics. Update: *A Report From the American Heart Association Statistics Committee and Stroke Statistics Subcommittee.*

[3]INML National Institute of Legal Medicine. 2008. *CRNV Reference Information Division Expert-Road Prevention Fund.*

[4]J. Edhouse, F. Morris. 2002. ABC of clinical electrocardiography: Broad complex tachycardia. Part I. *Br. Med. J. 324: 719-722*

[5]L. J. Colin. 2004. Electrocardiogram in the differential diagnosis of wide-QRS tachycardias. *Arch Cardiol Mex. 74:S44-S49.*

[6]O. Rueda. 2008. Programa de entrenamiento en electrocardiografía. *Libro Electrónico. 2ª. ed. Grupo de electrocardiografía, Universidad Industrial de Santander UIS. p 175.*

[7]Maron BJ, Roberts WC, Epstein SE. 1982. Sudden death in hypertrophic cardiomyopathy: a profile of 78 patients. *Circulation (7). 65:1388-94.*

[8]McKenna WJ, Camm J. 1989. Sudden death in hypertrophic cardiomyopathy: assessment of patients at high risk. *Circulation (5). 80:1489-92.*

[10]Krikler DM, Davies MJ, Rowland E, Goodwin JF, Evans RC, Shaw DB. 1980. Sudden death in hypertrophic cardiomyopathy: associated accessory atrio-ventricular pathways *B Heart J; 43 (3): 245-51.*

[11]Yetman AJ, McCrindle BW, MacDonald C, Freedom RM, Gow R. 1998. Myocardial bridging in children with hypertrophic cardiomyopathy. A risk factor for sudden death. *N Eng J Med; 339 (17): 1201-9.*

[12]Primo J, Geelen P, Brugada J, Filho AL, Mont L, Wellens F, et al. 1998. Hypertrophic cardiomyopathy: roles of the implantable cardioverter-defibrillator. *J Am Coll Cardiol; 31 (5): 1081-15*

[13]Elliott PM, Sharma S, Varnava A, Poloniecki J, Rowland E, McKenna WJ. 1999. Survival after cardiac arrest or sustained ventricular tachycardia in patients with hypertrophic cardiomyopathy. *J Am Coll Cardiol; 33 (6): 1596-601.*

[14]JA Álvarez-Fernández a, N Perales-Rodríguez de Viguri. 2005. International recommendations in resuscitation: from empirism to evidence based medicine. *Med Intensiva.29:342-8.*

[15]Akhtar M, Shenasa M, Jazayeri M, Caceres J, Tchou P. 1988. Wide QRS complex tachycardia, reappraisal of a common clinical problem. *Ann Intern Med; 109: 905-912.*

Journal of Physics: Conference Series **313** (2011) 012001 doi:10.1088/1742-6596/313/1/012001

[16]Brugada P, Brugada J, Mont L, Smeets J, Andries Ew. 1991. A new approach to the differential diagnosis of a regular tachycardia with a wide QRS complex. *Circulation; 83: 1649-1659.*
[17]Wellens HJ, Bar FW, Lie KI. 1978. The value of the electrocardiogram in the differential diagnosis of a tachycardia with a widened QRS complex. *Am J Med 64 (1): 27–33.*

System for the evaluation of power supply transformers in medical equipment

Conrado J Rodriguez, Carlos A Wevar Oller, Rodrigo G Bruni, Oscar R Vanella y Ricardo AM Taborda

L.I.A.D.E., Universidad Nacional de Córdoba, Argentina

liade@efn.uncor.edu

Abstract. The present work consists of the design and development of a prototype of equipment for the automatic testing of power supply transformers, following procedures and requirements of standard IRAM4220-1:2000 – IEC60601-1-1 "General requirements of safety for medical electrical equipment" [1], in order to fulfill certification of safety before commercialization. The device has a modular design thus facilitating modifications in case of updated requirements of standards or simply to improve performance. The equipment software, once the data of the transformer under testing is loaded, allows a friendly and simple interface with user while. It is worth mentioning that the equipment also contemplates the requirements of the international standard IEC 61010.1 (2) "Equipment for measurement, control, and laboratory use"

1. Introduction

The use of electrical energy in any device constitutes a potential risk for whoever makes contact with it. That is why diverse standards exist that establish methods to verify electrical safety, generally referred to certain field of application or group according to common characteristics and applications.

Electrical equipment for use in medical practice have the highest safety levels in comparison with almost every other application such as domestic appliances or office equipment. The reason behind such tough requirements is that they are often used in physical or electrical contact and/or transfer energy or substances to the patient, who may not be able to respond suitably to dangerous or painful conditions due to the absence of normal reactions (patient could be unconscious, anaesthetised, immobilized or simply have diminished physical capacities). Equipment designed for direct cardiac application have a direct connection to a vital organ so that a simple failure could lead to patient death.

It is worth saying that the safety of the operator might be jeopardized since even when he or she does not suffer any disease, in some cases the methods used by the devices represent a very high potential risk

The electrical safety of medical devices includes: safety of the equipment itself, of their installation in medical ambients and their application or use. It is required for normal use and also under the condition of first defect.

The reliability and quality of functioning is considered as an aspect of safety regarding vital assistance devices where the interruption of an examination or treatment is considered like a risk for the patient.

Standard IEC 60601-1-1, or its Argentine equivalent IRAM 4220-1, establishes the "general requirements of safety for AE" and details the corresponding tests to verify them.

From what has been described so far, we deduce that these devices must exhibit a high degree of reliability and quality of operation which need both to be verified before commercialization through quality certification.

An electro-medical equipment must have at least two independent means of protection against electric shock, both for operator and patient. There must always exist an element that provides safe isolation between parts under voltage, the accessible elements or that might enter in contact with who operates the equipment or the patient. In equipment class I, the second method of protection is the so called protection ground, meanwhile in those class II it is a second isolation known as additional.

The power transformer, besides its primary function of supplying the necessary voltages for the operation of the equipment, concentrates great part of the safety functions of the device, offering isolation between itself and the power network thru the correct insulation and separation of primary/secondary windings.

Thus it becomes essential to test its dielectric strength in order to verify integrity. Despite this, a fault in the equipment can originate an overload, and even a short circuit at the power supply area, causing a significant increase of temperature followed by degradation of the insulation.

For that reason it is necessary to determine the temperatures reached under overload and short circuit conditions, in order to evaluate if the materials used in the construction are apt.

Like in all systematic and repetitive process, its automatization allows simplification, thus diminishing the introduction of errors and the execution time.

For these reasons and following a line of work that the L.I.A.D.E (Applied Development and Research Lab), has pursued since long: development of equipment for testing biomedical equipment, is that we proposed the design, development and construction of a "SYSTEM OF EVALUATION OF OVERLOAD OF TRANSFORMING OF FEEDING FOR BIOMEDICAL EQUIPMENT" according to norm IEC 60601-1-1 /IRAM 4220-1.

2. Specifications

The testing system has the following specifications:

- Injection of a constant RMS current along the range 10mA – 5A +/- 5%, in transformers with secondary winding voltage range of 5 – 50V for a limit maximum power of 150W.
- Measurement of the winding coils resistance within the range 20mΩ to 200Ω with an accuracy better than 5%
- Calculation of the temperature of the windings
- Measurement of total testing time.
- Friendly software user interface
- Easy maintenance, in order to facilitate future modifications.
- Design according to standard IEC61010-1.

3. Development

In figure 1 we can observe that there exist four enslaved microcontrollers and four power stages, with their respective measurements of currents feedback, one for each secondary winding of the transformer, until a maximum of four, due to the adopted modularity principle.

Although the operating principle of the four is the same, we will describe the first one, because it has the control of the winding SBE (in overload) and the greatest current will circulate thru it, thus becoming the proper place to measure winding temperature. The others handle control of the current thru windings SCN and will only be mentioned whenever there are any differences.

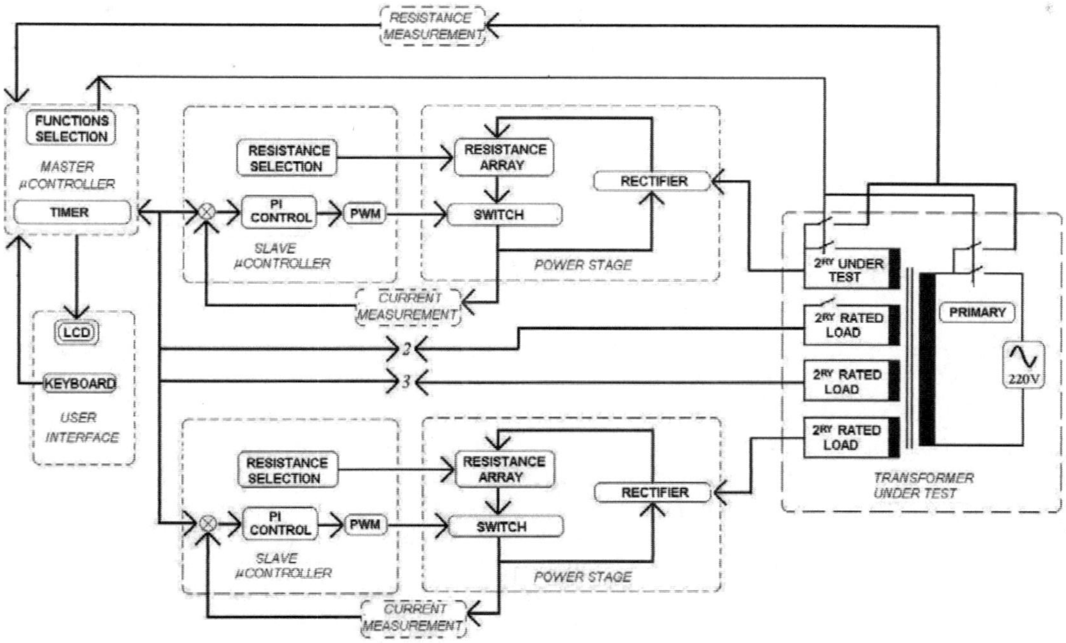

Figure 1. Block diagram of the evaluation system.

3.1. Master microcontroller

The key component of the equipment is a Microchip microcontroller PIC16F877A working with a 20MHz clock utilized to perform the following tasks:

- Function selection: measurement of windings resistance (primary and SBE) or loading the transformer under testing for current flow.
- Digitalization of the resistance measurement signal.
- Store samples of that signal.
- Analyze samples and perform the necessary calculations for determining windings temperature
- Store results
- Measure of total elapsed time during testing.
- User interface thru a 16 keys keyboard and a 4x20 LCD screen
- Data communication with the slave PIC´s thru I2C system.

As mentioned before, the system performs two main functions: measurement of resistances (primary winding and SBE) and circulation thru each winding of the transformer of a constant current.

Selection of functions is made by conmmuting three relays for the SBE case, as explained on table 1 and shown on figure 2.

Table 1. Relay commutation for function selection.

Relay1	Relay2	Relay3	Function
NC	NC	NC	Measure resistance of primary winding
NC	NC	NA	Measure resistance of SBE winding
NA	NA	X	Primary is connected to 220V and SBE is loaded for current flow test.

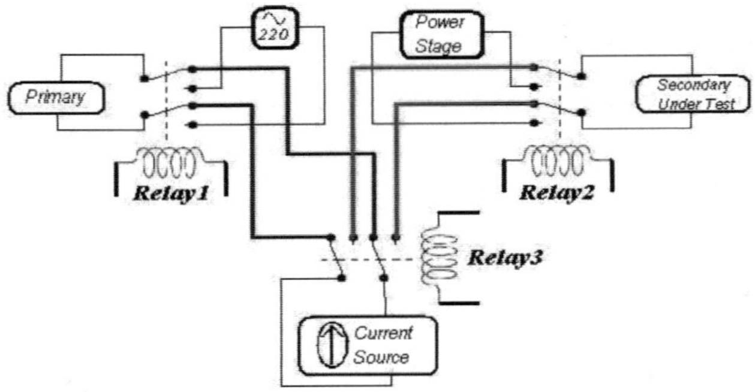

Figure 2. Selection of main functions.

Meanwhile in the SCN, the Slave PIC takes charge thru the PWM control and resistances bank selection.

For the digitalization of the resistance measurement, 50 data points are taken from the ADC and an average is then calculates. The digital analogic converter (ADC) works with 10 bits precision.

The variation of the resistance is utilized to calculate temperature. That is why the microcontroller samples winding resistance every 5 minutes.

$$T = Ti + \Delta t \tag{1}$$

$$\Delta t = \frac{(R\_medida - Ri)}{Ri} * (234.5 + Ti) \tag{2}$$

All the variables within the formula use 32 bits floating pint in the program thus achieving a greater resolution than specified for temperature.

The measurement of the total testing time uses an external clock of 32MHz in Timer 1.

For the communication, the Module I2C works as Master with a transmission speed of 100 kHz and I2C functions are forced by hardware. Data transfer is synchronous, that is the clock signal is sent to synchronize very bit in bit packets. La transferencia es sincrónica, es decir envía la señal de reloj para sincronizar cada bit, en paquetes de bits.

Each slave device connected to the bus is assigned a unique 7 bits coded address.

3.2. Slave microcontroller

A microcontroller PIC16F877A from Microchip, working with a 16 MHz clock is utilized to perform the following tasks:

- Receive and transfer data to the Master
- Digitalize the signal coming out of measurement of current flow thru secondary winding.
- Store samples of that signal
- Analyze and perform the necessary calculations to compare with data received by Master
- Provide a PWM signal by means of a PI control[3],[4] in order to carry out the control of load current.
- Select the appropriate resistance according to nominal voltage and current of the test.

A PI control was implemented, figure 3, because our process of current control does not require a time of fast establishment time, since the duration of the test is of 30 min. Therefore the derivational one is not necessary, but still we need as low an error as possible.

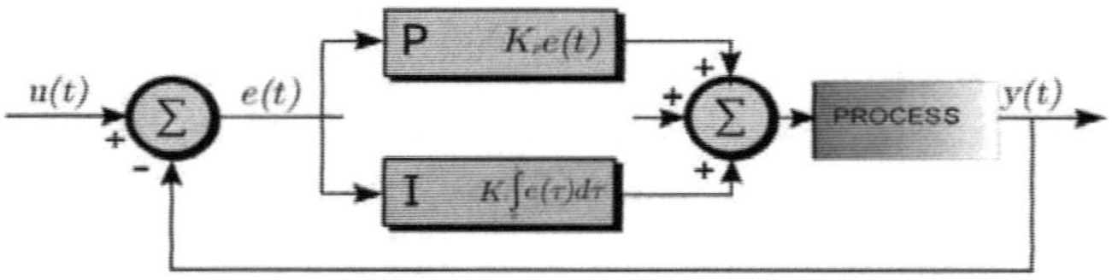

Figure 3. PI feedback control system.

$$\upsilon(t) = MV(t) = K_p e(t) + K_d \frac{de}{dt} \quad (3)$$

We assumed Ki=0 to tune controller. The gain Kp was increased until getting a proper response of the system to changes in setpoint without an excessive overshot or oscillations. It was fund Ki=0.1 to be an optimum value.

With this configuration the establishment time is achieved around 10 to 15 seconds depending upon setpoint, what is reasonably small as compared to total testing time.

The PI controller provides a PWM signal at 1 kHz with 10 bits resolution ranging from 0 to 100%, thus the control in this case has an output between 20 and 90% of the PWM.

The digitalization of the current measurement (RMS) was calculated with the discrete values formula:

$$I_{RMS} = \sqrt{\frac{1}{N}\sum_{i=1}^{N} x_i^2} = \sqrt{\frac{x_1^2 + x_2^2 + \ldots + x_N^2}{N}} \quad (4)$$

The converter acquires a total of 1000 samples of the signal, square each one and then adds them. The PIC configuration allows 170.4 μs sampling time. That means a time between samples of 171.4 μs or a sampling frequency of 5.8KHz.

Then the time needed for 1000 samples is 171,4 ms, that is, in one wave cycle (10 ms) pulsed at 1 kHz there are 58 samples, therefore fulfilling the sampling theorem.

The selection of resistance depends on voltage (V) and current (mA) data received from sampling. The program calculates the peak voltage of current flowing through the winding as:

$$Vpico = V(dato) * \sqrt{2} \quad (5)$$

Then the system determines the resistance to be selected simply with the Ohms law. Once determined, the closest value, always smaller than the calculated one, is selected. Otherwise the control would never reach the current preset.

3.3. Resistance measurement

A direct current supply pushes 25 mA current through the primary and SBE windings causing a voltage drop which is measured and signal-conditioned with an instrumentation quality differential amplifier with unit gain and then amplified with 4 possible pre-programmed values of gain automatically selected by the master PIC depending on signal amplitude.

It is worth noticing that if an alternate current were used instead, we would measure winding coil impedance instead of the pure resistance which is needed for temperature calculation.

3.4. Power stage

Here flows the current of the transformer under testing. We can partition this block in three for better understanding: rectifier circuitry, bank of load resistances and power commutation.

The rectifier circuit is simply a diode bridge that converts the sinusoidal alternating signal coming out of the transformer secondary into a full-wave rectified signal for further use.

The bank of load resistances is an array of 16 resistances for the SBE and 8 for the SCN, connected in parallel, that, when selected with help of the relays, will allow the flow of the proper current value according to the nominal voltage of the transformer.

The power switch[5] is made of a MOSFET transistor[6] and its corresponding snubber circuit for protection[7]. It will fix an average current according to the PWM signal injected in its gate which, in turn, will be measured and signal-conditioned by the following module.

3.5. Current measurement

This stage includes current sending and further signal conditioning. [8]

Current is detected by measuring the voltage drop across a shunt resistance in series with the MOSFET source. In the SBE case there exist two shunt resistances selected with relays meanwhile in the SCN, there is only one.

Voltage thus sensed is amplitude conditioned with a low noise differential amplifier which can be configured with either one of 5 gain values automatically selected by the slave PIC, depending on signal amplitude, with the help of a multiplexer.

This signal is then delivered to the A/D converter of the slave PIC for the calculation of the True RMS value.

3.6. User interface

It consists of a matrix keyboard 4x4, used to input nominal parameters, and a 4x20 LCD screen to visualize testing parameters and real values measured during testing and once finished.

4. Results

What follows is the presentations of results of performance tests of the equipment designed.

In order to realice them, 7 values were selected for each amplifying range of the instrument, as measured by a reference instrument, Hewlett Packard model 3478A for resistance and Fluke189 for current.

Those values were measured three times, in descending, ascending and descending order, respectively, in order to compare the average value with the reference one and thus obtain the percentage error, as shown in figures 4 and 5.

The lowest resistance values, between 150mohm and 20mohm, were constructed with a Nychrome wire of 1,3 ohm/meter, meanwhile commercial precision resistors were used for all other values. A 24V/6A transformer was used to excite input for the current measurement.

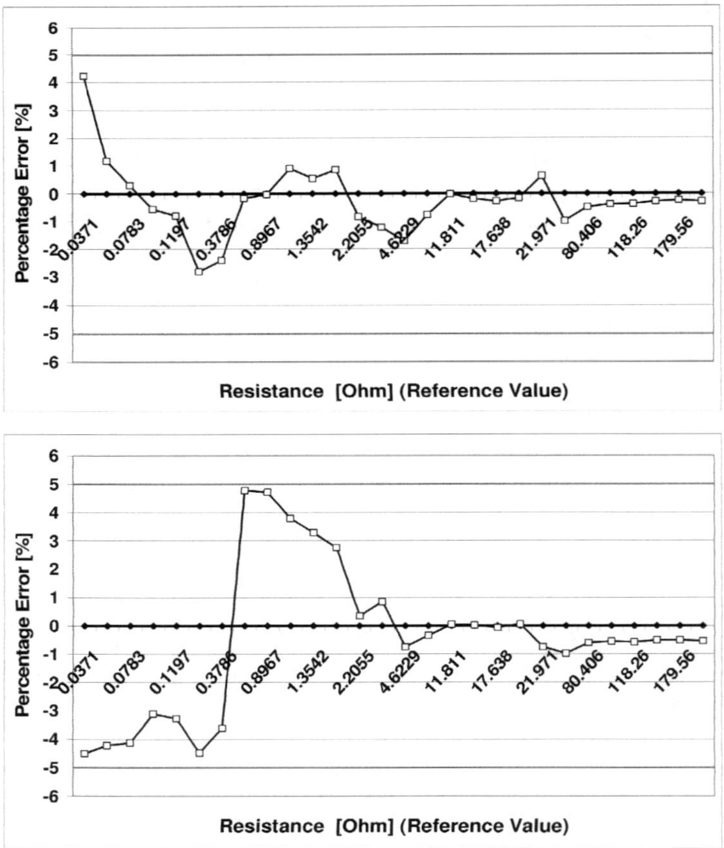

Figure 4. Percentage error in resistance measurement for primary and secondary circuits of transformer.

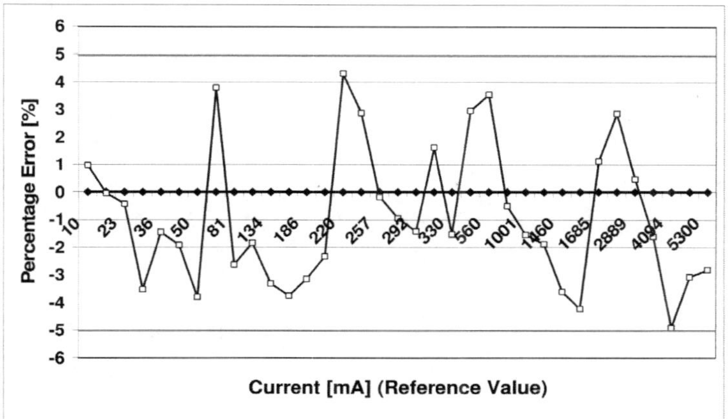

Figure 5. Percentage error in current measurement.

If we examine carefully the verifications realized by the equipment tester, we conclude that the mean error in measurements is smaller to the 5% required by the standards.

5. Conclusions

The accomplishment of a prototype that allows to perform the tests on supply transformers required by standard IEC 60601-1 and whose design also considers the requirements of the standard IEC 61010-1 for electrical safety of laboratory equipment.

Due to the principle of modularity employed, the equipment can be implemented partially according to particular requirements of transformers since during preliminary testing, every block was implemented and tested individually.

References

[1] IEC 60601-1:1996, 3ra Ed, Medical electrical equipment. Parte 1: General requirements for basic safety and essencial performance.
IRAM 4220-1, 2da Ed, Aparatos electrómedicos. Parte 1: Requisitos generales de seguridad
[2] UNE-EN 61010-1:2002 requisitos de seguridad de equipos eléctricos de medida control y uso en laboratorio. Parte 1: Requisitos generales.
[3] Benjamín Kuo- Ed Prince-Hall, "Sistemas de Control Automático" - 7ª Ed., 1996
[4] Katsuhiko Ogatta- Ed Prince Hall, "9. Ingeniería de Control Moderna- 3ª Ed., 1998
[5] M. H. RASHID, "Power Electronics: Circuits, Devices and Applications" – Ed Prentice-Hall, INC.- 1988.
[6] Motorola Power MOSFET Transistor Data.
[7] Keith H. Billings, C. Eng., M.I.E.E. "Switchmode Power Supply Handbook".
[8] J.G.Webster "Measurement, Instrumentation And Sensors Handbook" - Ed. Electronics – 1999.

Journal of Physics: Conference Series **313** (2011) 012003 doi:10.1088/1742-6596/313/1/012003

Management information system of medical equipment using mobile devices

C. Núñez[1], D. Castro[2]
[1] Biomedical Engineer, Universidad de Valparaíso.
[2] Biomedical Engineer, Universidad de Valparaíso.

nunezcarlos@live.cl

Abstract. The large numbers of technologies currently incorporated into mobile devices transform them into excellent tools for capture and to manage the information, because of the increasing computing power and storage that allow to add many miscellaneous applications. In order to obtain benefits of these technologies, in the biomedical engineering field, it was developed a mobile information system for medical equipment management. The central platform for the system it's a mobile phone, which by a connection with a web server, it's capable to send and receive information relative to any medical equipment. Decoding a type of barcodes, known as QR-Codes, the management process is simplified and improved. These barcodes identified the medical equipments in a database, when these codes are photographed and decoded with the mobile device, you can access to relevant information about the medical equipment in question. This Project in it's actual state is a basic support tool for the maintenance of medical equipment. It is also a modern alternative, competitive and economic in the actual market.

1. Introduction

In the last years there has been the phenomenon of global massification of mobile telephony. In South America the mobile telephony market it's in a continuous grow. Particularly in Chile the statistics indicates that nine of ten persons have a mobile device [1].

By increasing the supply in the market, the available technologies in this field are getting better, the mobile devices have greater capabilities and consequently it's possible to implement better applications in the devices. According to this, it raises the following question: How obtain benefits from these technologies in the biomedical engineering field?

Is in this context that the project is developed. Currently there are several management softwares for the medical equipment [2]. Unfortunately the current systems are expensive and inefficient in some cases.

In order to take advantage of current technologies and optimize the existing systems, from the point of view of work and cost, this project proposes to use mobile devices as a platform for a management information system of the medical equipment. The system manages to converge and to coexist: the information and communications technologies (through transmission), the digital image processing (by decoding barcodes with mobile devices), the programming languages and clinical engineering. All new and effective way.

2. Design

The system was designed as a mobile tool for managing information relevant to the departments of medical equipment.

Using a mobile phone, qr-codes [3] (2D barcodes) are decoded, which are associated with a medical equipment. The purpose of this decoding is to receive on the mobile device critical information (contained in a database) for the equipment in question (quick guide or checklists for example).

The use of qr-codes intended to facilitate the process of entering information on the mobile platform and make it more dynamic. Similarly, using the information associated with bar codes, the mobile platform is use to generate service orders related to the maintenance of biomedical equipment.

In this way then, the mobile device has the role of central axis of the system and through a Web server connectivity allows to obtain the requested information in all places. Figure 1 outlines the complete system and associated information flows.

Figure 1. Schematic of the system.

3. Development and Operation

In the development of the system, the main tools used and their characteristics are:

- Mobile Phone with Java support, 2.0 MP integrated camera and GPRS/EDGE connectivity.
- Apache Web server, with PHP and MySQL.

The system is based on the interaction between the software installed on the mobile phone and the managing databases from the Web Server.

The mobile device software was developed in Java for cell phones, better known as J2ME. The choice of this language is mainly due to high compatibility presenting with these devices.

The whole system, as shown in Figure 2, can be divided into the following stages:

a) 2D bar codes decode

This step involves accessing the mobile phone's integrated camera in order to photograph a qr-code and then decode it to obtain the associated information.

The data associated with the code mentioned may be the brand, model or service where is operating the medical equipment, among others. This decoding it simplifies the introduction of several parameters to the user, which are already in a database associated with the decoded barcode.

b) Send or request information
Once the qr-code it's decoded, is possible to perform various operations related to the maintenance of medical equipment. These options include:

- Generate and send a service order.
- Request for preventive maintenance checklist.
- Request for information relating to medical equipment in question (basic information, operation, last maintenance, photography, etc.).

c) Web Server Response
Depending on the code associated with the medical equip and the action required from the mobile device will be the answer that will run the Web server.

This response can be summarized as sending information from the system database or the inclusion of information in the database.

d) Reception of Information
Once is processed the response in the Web server, the information according to the action required by the mobile phone will be received and displayed.

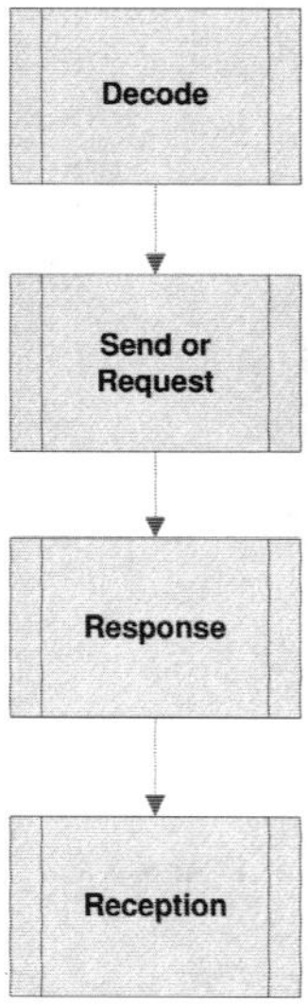

Figure 2. System stages.

4. Experimentation

The resulting product is a management tool based on the various stages that interact with each other, making a high-mobility system, easy access and easy use.

To determine the decoding capability of the system, using a medium category of mobile phone, several operational tests (called decoding tests) were conducted. In Figure 3 we can appreciate the mobile device photographing a QR-Code for further decoding.

The decoding tests consist in photographing bar codes by varying the size of the code (in [cm ^{2}]) and the distance from the camera to the code (in [cm]). 5 series are realized for each size and distance, thus quantifies the number of attempts required for a successful decoding. Finally is calculated the percentage of successful decoding for each series, according to equation (1):

$$\%Successful\_Decode = \frac{Successful\_attemps}{Attemps\_total} \bullet 100 \quad (1)$$

From the results of tests of decoding, is possible to infer minimum requirements for optimal system performance. The size of the qr-code must be at least 3.5 [cm] x 3.5 [cm] and the resolution of the integrated camera 1.3 MP or more. The above test results are presented in Table I:

Table I. Decoding tests results.

Format	Size [cms x cms]	Distance [cms]	Successful decode [%]
LCD Display	3,5 x 3,5	20	100
		20	100
Printed	6,5 x 6,5	10	100
		20	100
Printed	3,5 x 3,5	10	100
		20	60
Printed	2,2 x 2,2	10	80

Table II. Communication tests results.

Protocol	Decode time [seg]	Reception time [seg]	Total time [seg]
GPRS	2	20	22
Wi-fi	2	10	12
Wi-fi (emulated)	--	5	5

Whereas most stages of the system are directly related to the exchange of data between the mobile device and the Web server, communication tests were realized to the system. These communication tests consist in measuring the time it takes to get the information to the mobile device since the sending of a query.

The measured time considers the entire consultation process, since it decodes a qr-code until it reaches the requested information. The average size of information sent in these tests is 525 [bytes] and the server used is physically located in the United States (the mobile device was located in Chile). The test results are listed in Table II.

The results of both tests allow to state that from the point of view of time and cost, the system is fast and economical, making it highly competitive with any existing management software.

The estimated time it takes to the user to perform a query (according to the results of tests of communication) is about 25 seconds.

The associated cost is low because the system does not require of expensive platforms for operate. As demonstrated by tests carried out, a mid-class mobile device that meets the minimum requirements is enough to make the system work.

Figure 3. Mobile device capturing a barcode.

5. Discussion

The system is strongly influenced by the technology of mobile equipment used as a platform. The points where greater dependence was detected are those directly related to the technological features necessary for the application. These are the decoding of bar codes and data transmission.

Referring to the decoding, this depends directly on the resolution of the camera integrated into the mobile equipment. As shown by the results in Table I, with a mid-class mobile, the qr-codes are decoded without problems. With high-class mobile devices with integrated cameras of highest resolution (5MP for example) the decoding is optimized.

Relating to communication, the tests were realized using two different methods of data transmission, GPRS and Wi-Fi. Today mostly of the mobiles of medium category have the ability to communicate data through the GPRS protocol. This is an important factor to consider when implementing the system, since the mentioned protocol has some inherent limitations, for example the speed of transmission. For the project in its current state, the limitations of GPRS are not a problem, but are considerable to contemplate in a future evolution of the system.

On the other hand there is the use of Wi-Fi as data transmission protocol. The mobile devices with Wi-Fi are considered high-class and therefore its cost is higher, but the trend indicates that most mobile devices will have this technology in the near future [4], which will make more accessible to users.

Using Wi-Fi as a data transmission method brings many benefits compared to GPRS, mainly due to their speed of transmission. Yet, Wi-fi implies an investment cost when deploying the system in the sector (clinic or hospital) due to technological tools that are necessary (modems, routers, cabling, etc.) to allow connectivity.

In summary, the importance of mobile phone specifications are considerable, since items like a high quality camera, the connectivity or a high performance processor allows better use of the system.

Another relevant aspect of the project is related to its safety, here are two angles, the safe use of mobile devices in hospitals and safe handling of information.

About the use of mobile devices in hospital compounds, exist a controversy about this subject. There are studies that claim that cell phones generate significant electromagnetic interference on some medical equipment [5], and there are many others who deny this information [6]. But there are not conclusive studies about any of these two statements.

In order to reconcile any position regarding the above, a guideline for a safe use is considered, which is *"not using the application in medical devices that are connected directly or indirectly to a patient."* Compliance with this guideline does not generate conflict for the system, because the focus of this project is the maintenance management and, by definition, medical equipment in maintenance must be disconnected from any patient.

As regards to the safe handling of information, to have a reliable system, it is necessary to restrict the access to the database (for example: using passwords) and back up the information continuously. It is also necessary to identify the mobile phone from which is requesting or sending information, in order to restrict access only to authorized persons.

6. Conclusion

The department of medical equipment from any health institution, public or private, plays a critical and important role, because they are responsible for ensuring that all medical equipment from the organization will have a correct and efficient performance and provide a quality service.

To assist the staff of the medical equipments departments, there are management or administration softwares, which facilitate the several tasks.

The implemented system emerges as an alternative to the existing management softwares, this system optimizes the processes related to the maintenance of medical equipment. Using the mobile devices as a platform, the application becomes versatile, dynamic, universal and flexible.

Adding new functions to the system is a real possibility and that is already developing thanks to the continued advancement of mobile technologies and the great potential that shows the application.

The project in its current state is the foundation of a complete management application, competitive and affordable for the departments of medical equipment of any health institution (public or private). It is also an alternative for the assets control of the maintenance enterprises of medical equipment.

Acknowledgments

To our teachers Pablo Roncangliolo and Cesar Galindo for being our mentors and guides in this project. To our friend Rafael Zuaznabar for his support and for their future participation in the project. To our families for supporting us during the long process of training as a biomedical engineer.

References

[1] Subsecretaría de Telecomunicaciones (Subtel). Available in: http://www.entelpcs.cl/noticias/

[2] "Datastream lanzo MPX Mantec, una nueva solución para la administración de del ciclo de vida de los activos", Revista electrónica Mazalan Comunicaciones, September 2001. Available in: http://www.mazalan.com

[3] Denso-Wave, QR_Codes. Available in: http://www.denso-wave.com/qrcode/index-e.html

[4] Andrea Kobylnik. "Nueva generación de equipos móviles", November 15, 2007. Available in: http://www.tendenciasdigitales.com.ar/

[5] Cheryl I. Shaw; Robert M. Kacmarek; Rickey L. Hampton; Vincent Riggi; Ashraf El Masry; Jeffrey B. Cooper; William E. Hurford, "Cellular phone interference with the operation of mechanical ventilators". Crit Care Med 2004 Vol. 32, No. 4, pp 928 - 931.

[6] Reachel C. Vreeman and Aaron E. Canoll, ""Medical myths: Mobile phones are dangerous in hospitals", BMJ 22-29 December 2007, Volume 335, pp1289. Available in: http://bmj.com/cgi/content/full/335/7633/1288

Patellofemoral pain syndrome: electromyography in a frequency domain analysis

D S Catelli[123], H U Kuriki[12], L F Polito[12], F M Azevedo[2], R F Negrão Filho[2] and N Alves[2]

[1] Post-Graduation Program Interunits in Bioengineering, EESC/IQSC/FMRP-USP, Av. Trabalhador São-carlense, 400, 13566-590, São Carlos/SP, Brazil
[2] Laboratory of Physical Therapy Applied to Human Moviment, FCT-Unesp, Rua Roberto Simonsen, 305, 19060-900, Presidente Prudente/SP, Brazil

catelli@sc.usp.br

Abstract. The Patellofemoral Pain Syndrome (PFPS), has a multifactorial etiology and affects approximately 7 to 15% of the population, mostly women, youth, adults and active persons. PFPS causes anterior or retropatelar pain that is exacerbated during functional motor gestures, such as up and down stairs or spending long periods of time sitting, squatting or kneeling. As the diagnostic evaluation of this syndrome is still indirect, different mechanisms and methodologies try to make a classification that distinguishes patients with PFPS in relation to asymptomatic. Thereby, the purpose of this investigation was to determine the characteristics of the electromyographic (EMG) signal in the frequency domain of the *vastus medialis oblique* (VMO) and *vastus lateralis* (VL) in patients with PFPS, during the ascent of stairs. 33 young women (22 control group and 11 PFPS group), were evaluated by EMG during ascent of stairs. The VMO mean power frequency (MPF) and the VL frequency 95% (F95) were lower in symptomatic individuals. This may be related to the difference in muscle recruitment strategy exerted by each muscle in the PFPS group compared to the control group.

1. Introduction

The patellofemoral pain syndrome (PFPS) is characterized as an anterior or retropatelar pain in the absence of other disorders and presents clinically as a condition of diffuse pain exacerbated by activities such as climbing stairs, getting a lot of time sitting, squatting and kneeling [1]. The PFPS is a common disease in athletes and affects about 7 to 15% of the population of young active adults, being more common in women, and one of its features is a change in the patellofemoral joint from a patellar instability, generating a slope or lateralization of the patella [2]-[4]. Based on the concept of balance of forces involved in this joint, the quadriceps muscles - *vastus medialis obliquus* (VMO), *vastus lateralis* (VL), *rectus femoris* (RF) and the *vastus intermedius* - would act together to a uniform extent of the knee, and that lateralization of the patella could be generated by any imbalance in the presence of these forces, many studies have used electromyography (EMG) as a method to describe these muscles in different motor performances [5]-[7]. The EMG is applicably to the surface muscles (VMO, VL and RF) and could be useful clinically to diagnose individuals with the syndrome, or even to characterize and classify individuals likely to develop the disease.

[3] To whom any correspondence should be addressed.

Journal of Physics: Conference Series **313** (2011) 012004 doi:10.1088/1742-6596/313/1/012004

Trying to make this classification, studies involving analysis of the EMG signal in time domain have been widely used, primarily to diagnose the early activation of the VMO/VL and their relationships. However, there are still many contradictions surrounding this aspect [8]-[11]. The analysis of the EMG signal in the frequency domain have proven effective with respect to criteria for designation of characteristics and types of muscle contraction, as well as determination of muscular fatigue [12]-[14]. For the frequency analysis, the signal should has specific features: stationarity, trend and seasonality. The analysis parameters commonly used are the mean power frequency (MPF), that is the point which divides the area below the frequency spectrum into two equal parts; the 95% frequency (F95), a point that focuses on your left 95% of the total area below the spectrum; and the peak frequency (Fp), which determines the peak frequency in the spectrum.

Thus, the purpose of this study was to determine the frequency characteristics of the EMG signal in patients with PFPS during the weight transfer in ascent of stairs.

2. Materials and methods

2.1. Subjects

33 volunteers were submitted to the EMG analysis. The individuals were divided into two groups: 22 in the control group (mean ± SD: age 23 ± 2 years, body mass 55 ± 6 kg, height 161.2 ± 4.8 cm) and 11 clinically diagnosed with PFPS (age 24 ± 3 years, body mass 57.3 ± 5.3 kg, height 161.9 ± 5.9 cm). The protocol for diagnosis PFPS consisted of tests on the knee joint to exclude signs of the syndrome or other changes in the patellofemoral joint. Prior to participation, all procedures were explained and participants signed an informed consent for the study. The study was approved by the Ethics Committee in research of the FCT / UNESP, n.166/2007 process.

2.2. Exercise protocol

The experimental protocol was the rise of three steps (12 replicates) with 20 cm in height, in a continuous and alternating way. During the course of this experiment were collected signals of *vastus medialis oblique* (VMO) and *vastus lateralis* (VL) of the right lower limb, that were the dominant one for the control volunteers and the painful member of the PFPS group volunteers.

2.3. Measurements

For the assessment it was used two pairs of Ag/AgCl surface electrodes, 3M® Meditrace® model, with a 10mm diameter captation surface. The electrodes were placed in parallel, with 20 mm of distance between each other. A pre amplification circuit was used, with a gain of 20 times, CMRR (Common Mode Rejection Ratio) greater than 80 dB and an impadance of 1012 Ω. The signals were captured in a signal conditioner module, LYNX® BIO EMG 1000 model, using three channels for EMG signals. The signals were filtered with a digital band-pass filter of 20 – 500 Hz. All channels had a final gain of 1000 times and 4000 Hz of sampling frequency. The acquisition and storage of the signals were made using the LYNX® Bioinspector 1.8 software.

2.4. Data processing

The signals had to have the prerequisites to the Fourier Transformed calculation (stationarity, seasonality and trend). Therefore, the EMG peak activation was determined (by linear envelopment) and then it was selected 250 points before and after it; determining, in this way, 501 points or 125 ms, that was extracted to analysis. All signals were processed in MatLab®, with an algorithm made to attain MPF and F95 of the signals.

2.5. Statistical analysis

A Shapiro-Wilks normality test was applied. After the normality detection of the data, a t-student statistical test, for independent samples was applied, with a significance level of 5%. The tests were applied comparing both, the groups and the muscles inside the same group.

3. Results

For a better visualization of the means and respective standard deviations, they are disposed in Table I.

Table 1. Final results of the Mean power frequency of each muscle, according to given group.

	Control (Hz)	PFPS (Hz)	p-value
MPF – VMO	61.59±10.7	58.15±8.6	0.004[a]
MPF – VL	60.24±9.8	58.96±9.1	0.251
p-value	0.135	0.515	

[a] refers to statistical significant difference.

Table 2. Final results of 95% frequencies of each muscle, according to given group.

	Control (Hz)	PFPS (Hz)	p-value
F95 – VMO	160.60±34.9	162.51±37.7	0.649
F95 – VL	154.04±30.9	144.99±26.8	0.011[a]
p-value	0.025[a]	0.0002[a]	

[a] refers to statistical significant difference.

Looking at the MPF analysis, there was a statistically significant difference (p=0.004) between control and PFPS groups, in the VMO muscle EMG signal. For the F95 analysis, a statistical significant difference (p=0.011) was verified between the groups in the VL muscle. Still in the F95 analysis, it was observed a difference between the VMO and the VL muscles, inside the PFPS (0.0002) and the control (p=0.025) groups. Nevertheless, the lowest power index in the test was found in the last case (power = 0.61), it being known that the closest to 1 shows a more reliability.

The graphics of the Figures 1 and 2 show an ilustrative comparison of the frequencies and where were founded the statistical differences.

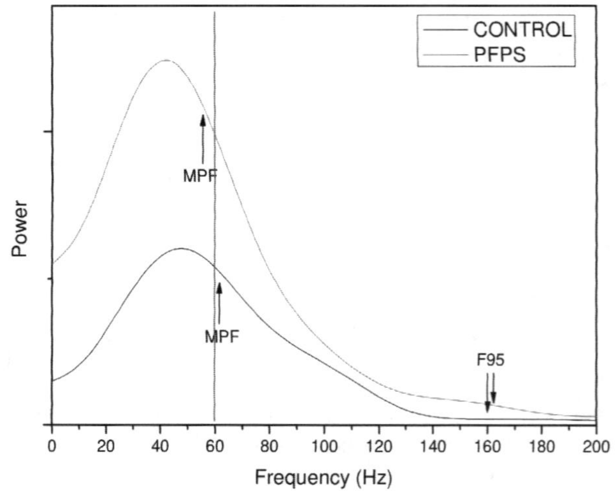

Figure 1. Illustration of the power spectrum in the VMO muscle in both groups with the blue line indicating where there was statistical difference.

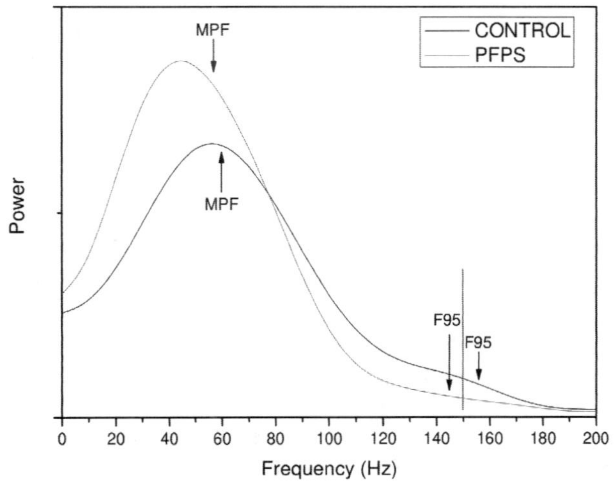

Figure 2. Illustration of the frequency spectrum in the VL muscle in both groups with the blue line indicates where there was statistical difference.

4. Discussion

As previously related, the aim of this paper was to define the characteristics of the power spectrum of an EMG signal during weight shift phase in stair climbing, in a group of patients with PFPS. Originally, the hypothesis was to find a muscular recruitment strategy difference between VMO and VL muscles, for symptomatic individuals, when compared to individuals of a control group.

The mean power frequency represents a central tendency mesure, that divides in half the total area of the spectrum. The F95 is a point where 95% of the total spectrum area is concentrated. Together, both provided important information of the spectrum morphology in relation to data compression in the plot.

Comparing specifically each muscle, it could be observed that in the VMO muscle, the centrality pattern (MPF) is distinct in each group. However, the same was not found when looking to F95, showing, thus, a greater data dispersion in the PFPS group, mainly in the final half of the graphic. In the VL muscle, this relationship was inverted, as it was not possible to find differences in MPF, but in F95, showing a greater dispersion in the control group data. A greater dispersion in the frequency data plot may denote recruitment of muscle fiber with distinct contraction characteristics [13, 15, 16]. It is evident that a slight imbalance of those characteristics between the groups exists, when making a comparison of the relationship of each muscle performance, because the VL muscle data, in the PFPS group, is more concentrated that in the control group, and the VMO muscle data is more disperse.

When comparing the muscles, similar morphologic characteristics on the graphics were found in both groups. It was showed an equalitie of the MPF between VMO and VL muscles, and a more evident dispersion in the VMO plot related to F95 in the VL muscle. However, it has to be emphasized that, in the control group, this difference found in F95 (p=0,025) do not has the same impact that the difference found in the PFPS group (p=0,0002), what indicates a more evident data dispersion in the symptomatic group.

5. Conclusions

The opposite density distribution of the VMO and VL muscles, when the MPF and F95 of the control group is observed, compared to the PFPS group, can represent a distinct relationship in the muscle fibers recruitment tendency. However a normalized analysis of the power spectrum characteristics, as well as the peak frequency, could be helpful to reach an even tighter conclusion about the parameters and characteristics of a frequency domain analysis in patients with patellofemoral pain syndrome. An EMG analysis of others motor gestures could be very useful to complement this study.

Acknowledgements
The authors thank CAPES and FAPESP (2008/02194-9) for the financial support.

References

[1] S.M. Cowan; P.W. Hodges; K.L. Bennell; K.M. Crossley 2002 Altered vastii recruitment when people with patellofemoral pain syndrome complete a postural task *Arch Phys Med Rehabil* **33** 989-95
[2] K.M. Crossley; S.M. Cowan; K.L. Bennell; J. Mcconnell 2004 Knee flexion during stair ambulation in individuals with patellofemoral pain *Journal of Orthopaedic Research* **22** 267-274
[3] J. Laprade; E. Culham; B. Brouwer 1998 Comparison of five isometric exercises in the recruitment of the vastus medialis oblique in persons with and without patellofemoral pain syndrome *J. Orthop. Sports Phys. Ther.* **27-3** 197-204
[4] R.P. Csintalan; M.M. Schulz; J. Woo; P.J. Mcmahon; T.Q. Lee 2002 Gender differences in patellofemoral joint biomechanics. *Clinical Orhopaedics and Related Research* **1-402** 260-269
[5] S.M. Cowan, K.M. Crossley 2009 Does gender influence neuromotor control of the knee and hip? *Journal of Electromyography and Kinesiology* **19** 276-282
[6] A.K. Stensdotter, H.Grip, P.W. Hodges, C. Häger-Ross 2008 Quadriceps activity and movement reactions in response to unpredictable sagittal support-surface translations in women with patellofemoral pain. *Journal of Electromyography and Kinesiology* **18** 298-307
[7] K. Cerny 1995 Vastus medialis oblique / vastus lateralis muscle activity ratios for selected exercises in persons with and without patellofemoral syndrome *Phys. Ther.* **75-8** 672-83
[8] D. Bevilaqua-Grossi; L.R. Felicio; R. Simões; K.R.R. Coqueiro; V. Monteiro-Pedro 2005 Avaliação eletromiográfica dos músculos estabilizadores da patela durante exercício isométrico de agachamento em indivíduos com síndrome da dor fêmoropatelar *Revista brasileira de medicina do esporte* **11-3** 159-163
[9] M.L. Voight; D.L. Wieder 1991 Comparative reflex response time of vastus medialis obliquus and vastus lateralis in normal subjects and subjects with extensor mechanism dysfunction. *Am J. Sports Med.* **19-2** 131-137
[10] S.M. Cowan; K.L. Bennell; K.M. Crossley; P.W. Hodges; J. Connell 2002 Physical therapy alters recruitment of the vasti in patellofemoral pain syndrome *Medicine & Science in Sports & Exercise* **34-12** 1879-85
[11] H.U. Kuriki 2009 *Análise de parâmetros biomecânicos relacionados à Síndrome Dolorosa Fêmoro-Patelar* (São Carlos:São Paulo University – Bioengineering)
[12] P.A. Kaplanis; C.S. Pattichis; L.J. Hadjileontiadis; V.C. Roberts 2009 Surface EMG analysis on normal subjects based on isometric voluntary contraction *Journal of Electromyography and Kinesiology* **19** 157-171
[13] M. Cifrek; V. Medved; S. Tonkovic; S. Ostojic 2009 Surface EMG based muscle fatigue evaluation in biomechanics *Clinical Biomechanics* **24** 327–340
[14] A.S.C. Oliveira; M. Gonçalves 2009 EMG amplitude and frequency parameters of muscular activity: Effect of resistance training based on electromyographic fatigue threshold *Journal of Electromyography and Kinesiology* **19** 295-303
[15] E. Kwatny; D.H. Thomas; H.G. Kwatny 1970 An application of signal processingtechniques to the study of myoelectric signals *IEEE Trans. Biomed. Eng.* **17-4** 303–313

[16] T.W. Beck; T.J. Housh; G.O. Johnson; et al 2007 Does the frequency content of the surface mechanomyographic signal reflect motor unit firing rates? A brief review *Journal of Electromyography and Kinesiology* **17-1** 1-13

Journal of Physics: Conference Series **313** (2011) 012005 doi:10.1088/1742-6596/313/1/012005

Histomorphological evaluation of Compound bone of Granulated Ricinus in bone regeneration in rabbits

Christiano Pavan Mateus, Gilberto Orivaldo Chierice, Tetuo Okamoto

Interunits Program in Bioengineering, EESC, FMRP, IQSC, University of São Paulo, São Carlos, USP

christiano@sc.usp.br

Abstract. Histological evaluation is an effective method in the behavioral description of the qualitative and quantitative implanted materials. The research validated the performance of Compound bone of Granulated Ricinus on bone regeneration with the histomorphological analysis results. Were selected 30 rabbits, females, divided into 3 groups of 10 animals (G1, G2, G3) with a postoperative time of 45, 70 and 120 days respectively. Each animal is undergone 2 bone lesions in the ilium, one implemented in the material: Compound bone of Granulated Ricinus and the other for control. After the euthanasia, the iliac bone was removed, identified and subjected to histological procedure. The evaluation histological, histomorphological results were interpreted and described by quantitative and qualitative analysis based facts verified in the three experimental groups evaluating the rate of absorption of the material in the tissue regeneration, based on the neo-bone formation. The histomorphologic results classified as a material biocompatible and biologically active. Action in regeneration by bone resorption occurs slowly and gradually. Knowing the time and rate of absorption and neo-formation bone biomaterial, which can be determined in the bone segment applicable in the clinical surgical area.

1. Introduction

Currentlly, the amount of materials used in the repair of bone discontinuity has increased.

In search of a material to be used in the correction of a bone discontinuity, ie to promote osteoconduction and / or osteoinduction and therefore reparative osteogenesis, autologous bone is the most appropriate indication [1].

Parallel to its availability, a replacement or an ideal bone graft needs to be found in the amount and form required to be deposited in the bone defect [2].

Osteoconductive biomaterials possess properties that serves as a skeleton or frame for the arrival of cells, where they will host. Leading to cell proliferation and osteoblastic activity [3],[4].

The biomaterials are used in medical devices, especially those who are temporarily or permanently implanted in the human body. The term biomaterial was defined at the conference of the National Institute for Health Development in 1982 as "any substance (no drug) or combination of substances, natural or synthetic origin, that can be used for a period of time, completely or partially as part of a system that treats, augment or replace any tissue, organ or body function"[5].

The application of biomaterials in surgical field – dental and orthopedic treatment, is focused on generation properties of a material with perfect reproducibility of bone, with the biocompatibility, biodegradability, the ability to initiate osteogenesis, composition and mechanical properties similar to those bone [6].

The biomaterial addition to their specific mechanical function once implanted, is a needed for compatibility with the properties of living tissue. These materials are divided into four groups: metals, ceramics, polymers and composites, taking into account that according to the induced biological response can be classified as: biotolerables, bioinert, bioactive and biodegradable [7].

The biomaterial contact with living tissues through the interface will influence a cellular response / tissue, thus activating the desired mechanism and obtaining accurate results [8].

Bioinert and biotolerables materials and leagues such as titanium and carbon, alumina (alpha-Al2O3) and zirconium oxide (ZrO2) are in response of bone tissue encapsulation of the implant by a layer of fibrous tissue [9].

Biodegradable materials: calcium sulfate and calcium triphosphate are designed to degrade or be engulfed by the body gradually, allowing a complete replacement of the implant to new tissue growth [9].

The material to induce a specific biological response in living tissue, forming a bond between the tissue and the implanted material (osteointegration), without the intervention of a layer of fibrous tissue, taken for the bioactivity [9].

The use of chemicals to facilitate bone regeneration through osteogenesis, osteoinduction and osteoconduction, is the target of ongoing research [10], [11].

And according to [12], it includes the factors of biocompatibility, absorption, bioactivity, biodegradation, depending on the physical and chemical properties of the material, which must be compatible with the physiological reactions of bone.

The compound bone of granulated Ricinus (Ricinus communis) when implanted in biological tissue has chemical stability, mechanical strength and biocompatibility [13], increasingly in the field of bone tissue restoration.

The study monitored by the bone regeneration biomaterial, based on new bone formation by histological evaluation.

2. Materials and methods

The experiments of this work were conducted with 30 female rabbits (Oryctolagus cuniculus), adult age of 90 days with weights between 2.5 and 3.0 kg.

The rabbits were divided randomly into 3 experimental groups of 10 animals (G1, G2 and G3), establishing post-operative periods of 45 days for G1, 70 days for G2 and 120 days for G3. Each animal received 2 bone defects in the ilium, one implemented em material defects and the other for control.

The implanted material was:

> Compound bone of Granulated Ricinus (COR)* in presentation granulated with 450μm grain size, being a polymer derived from castor oil (Ricinus communis).
> * Poliquil Araraquara Polímeros Químicos Ltda.

In the pre-operative procedures, the animals are exposed to fast for 8 hours and is performed trichotomy in the region to be operated.

It applies preemptive analgesia with morphine sulfate in doses of 1mg. kg every 6 hours for 24 hours (intramuscularly) and cloridrate Tramadol in doses of 2 mg.kg-1 every 8 hours for 72 hours (intramuscularly).

Like antibiotic, enrofloxacin is supplied by 1 mg.kg-1 every 24 hours for 7 days (intramuscularly). Protocol dissociative anesthetic agents with Xylazine and tiletamine-zolazepam at a dose of 5 mg.kg-1 and 15mg.kg-1, respectively (intramuscularly).

Then, antisepsis with chlorhexidine digluconate to 4%. The surgical procedure of the study region, the iliac bone is made by cranio-dorsal approach. An incision in the skin and subcutaneous tissue, and divided the origin of the fascia of the gluteus medius muscle to provide access and elevation separation subperiostal.

With a trephine of 4 mm external diameter attached to a micro motor for low speed, we made two bone defects in iliac bone under constant irrigation with saline solution 0.9%. A bone defect was filled with Compound bone of Granulated Ricinus and the other for control.

After filling the bone defects, the starts in continuous pattern suture to the gluteus medius muscle with absorbable thread and suturing of skin and subcutaneous tissue with thread no absorbable in simple interrupted pattern.

Surgical times are recorded in photographs.

After the procedures, the rabbits were brought to their cages, fully recovered, not causing weight loss and keeping intact surgical wounds.

After the experimental period of the 3 groups, practiced euthanasia of animals under ethical criteria.

The surgical specimens were removed, identified by a separate group and placed in containers with 10% neutral formalin.

The pieces were removed for histology laboratory for histological processing stages. Were made semi-serial cuts of 6 μm in thickness longitudinally on a rotary microtome. Mounted on slides and stained using the technique of hematoxylin-eosin (HE).

Slides were analyzed and photographed by Zeiss photomicroscope.

The selection of slides for the histological evalution, is attributed to interactions of the interfaces bone / implant.

Histomorphologic results are interpreted and presented by descriptive analysis based on the interplay of material in the three postoperative periods, evaluating the rate of resorption in tissue regeneration based on the newly boné formation.

The study monitored by the bone regeneration biomaterial, based on new bone formation by histological evaluation.

3. Results

In the experimental group G1, surveying large areas occupied by compound bone of granulated Ricinus and the formation of small bone spicules between them (Fig. 1A). In other slides, most notably the amount of newly formed bone tissue filling the spaces between fragments of the material (Fig. 1B).

Besides newly formed bone tissue, is evident in some areas the presence of connective tissue without bone differentiation (Fig.1C).

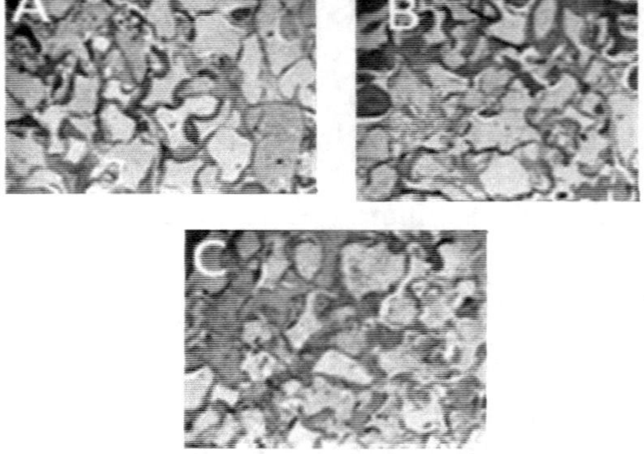

Fig.1. A- Small bone spicules; **B-** Greater amount of newly formed bone tissue; **C-** Connective tissue without bone differentiation. H.E., 25x.

In the control group G1, the surgical area is occupied by thin bone trabecula and wide space intertrabecular (Fig. 2A). In other slides, the bone trabeculae are more crowded and occupied by connective tissue between the bone trabeculae (Fig. 2B).

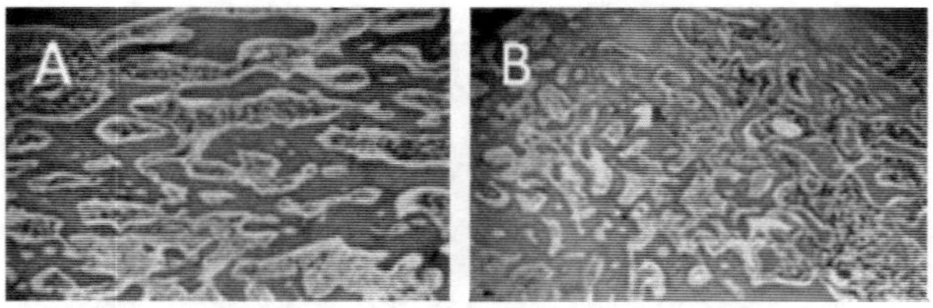

Fig.2. A- Surgical areal occupied by thin trabeculae; **B-** Connective tissue between the bone trabeculae. H.E., 25x.

Can be observed in G2, an extensive surgical area occupied by the compound bone of granulated Ricinus (Fig.3A), and note on the extremities, connective tissue without bone differentiation with a moderate number of fibroblastos (Fig. 3B). The other slides, there is presence of trabecular bone more developed and less of the material (Fig. 3C).

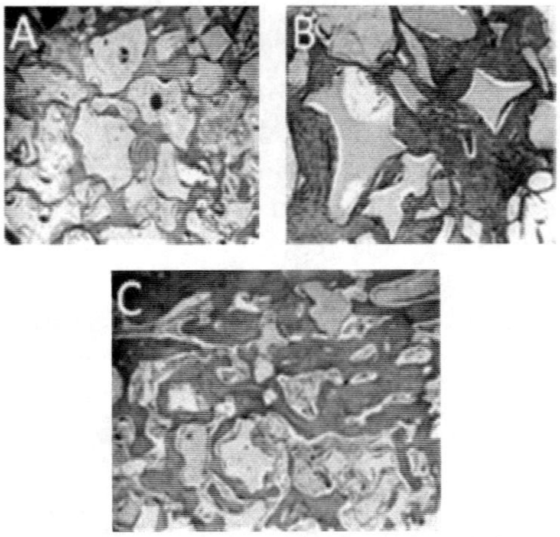

Fig.3. A- Specimen showing large area of the surgical space, occupied by the material, **B-** same specimen in Fig. 3A, showing areas with connective tissue without bone differentiation, **C-** bone trabeculae more developed and less material. HE., 25x, 63x, 25x.

All surgical areas in the control group G2 are occupied by thin bone trabeculae and found large areas with connective tissue without bone differentiation (Fig. 4).

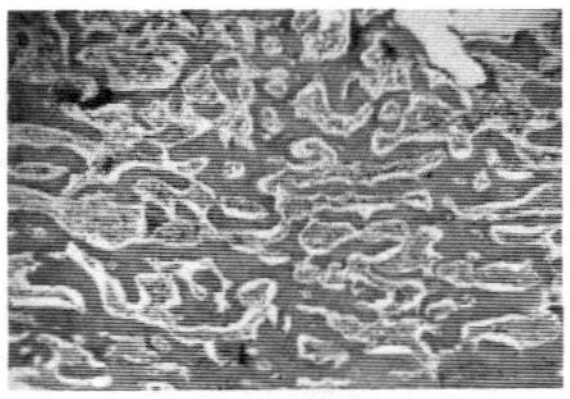

Fig.4. Thin bone trabeculae filling the surgical area. HE., 25x.

The observation period in G3, the bone defect where the compound bone of granulated Ricinus was deposited, there was a greater quantity of material and note discrete areas occupied by bone tissue (Fig. 5A). In one of the slides, is shown moderate amount of material next to trabeculae of crowded (Fig. 5B).

In other cases, there is the presence of lower quantity of material and the presence of newly formed bone trabeculae in contact with the polymer (Fig. 5C).

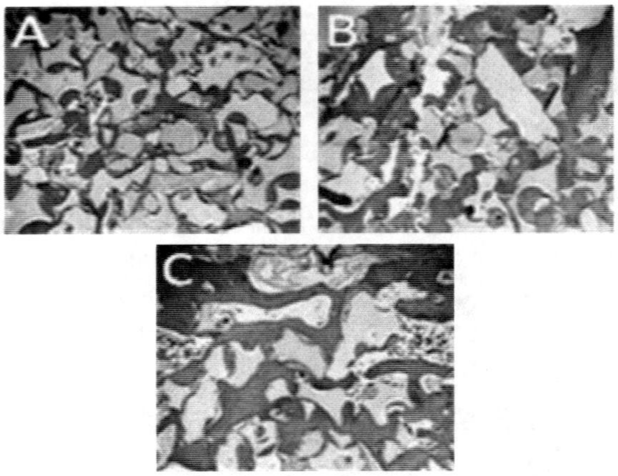

Fig.5. A- Specimen showing greater amout of material with discrete areas occupied by bone tissue; **B–** Moderate amount of material next to thicker trabeculas;
C– Presence of newly formed bone trabeculae in contact with the polymer. H.E., 25x.

In the control group G3, in all slides, the surgical area is fully proven by well developed bone tissue (Fig. 6).

Fig. 6. In all slides, the surgical area is occupied by trabecualr bone well developed. H.E., 25x.

4. Discussion

The histological evaluation determined that for longer periods, the implanted materials suffered a slow and gradual resorption

It also coincides with the position taken by [14] on slow resorption, defining that the use of um material that is slowly absorbed, with strength and elasticity close to the bone and capable of stimulating bone regeneration, would be revolutionary in the filling of bone defects and in the manufacture of implants.

This would also eliminate the disadvantages of autologous graft and metal implants.

The histological evaluation validates the its application to be used to evaluate the biocompatibility of implanted material, as cited by [15], where most of the protocols are based on the qualitative determination of relative number of cell types and quantity components of the extracellular matrix around the implants.

The performance of compound bone of granulated Ricinus in the three groups is shown surrounded by a developing trabecular bone, which achieves its physical form having an area with a high number of granules, promoting cell migration and obtaining a matrix bone.

The findings come from meeting with [16], reporting that osteoblasts prefer pore size ranging from 200 to 400 μm in diameter to facilitate migration, adhesion and proliferation.

The Compound bone of Granulated Ricinus has the approval from the Food and Drug Administration (FDA), U.S. government agency responsible for authorizing the circulation of new foods and medicines, in June 2003.

5. Conclusions

The new bone formation was characterized by a slow and gradual resorption.

The Compound bone of granulated Ricinus proved to be biocompatible and biologically active.

The days of G1, G2 and G3 will be the basis for understanding the rate of resorption and new bone formation of Compound bone of granulated Ricinus, can determine which (is) segment (s) of bone (s) is applicable within the clinical area -surgical. sfdsdfs

6. Future work

Put under study the evaluation by computer analysis of qualitative and quantitative bone mineral density, wich can be monitored the bone regeneration of the material.

References

[1] Tsonis PA – Regenerative biology: the emerging field of tissue repair and restoration. Differentiation, 70(8): 397-409, 2002.

[2] Zheng L, Wang Q – The current situation and future of extracellular matrix materials bone tissue engineering. Sheng Wu Yi Xue Gong Cheng Xue Za Zhi. 18(3):470-474, 2001.

[3] Boeck-neto, R.J.; Gabrielli, M.; Lia, R.; *Marcantonio, E.; Shibli, J.A.; Marcanto*-nio, E.J.R. Histomorphometrical analysis of bone formed after maxillary sinus floor augmentation by grafting with a combination of autogenous bone and demineralized freeze-dried bone allograft or hydroxyapatite. J Periodontol, v.73, p.266-70, 2002.

[4] Artzi. Z; Nemcovsky, C.E; Dayan, D. Nonceramic hydroxyapatite bone derivative in si-nus augmentation procedures: clinical and histomorphometric observations in 10 con-secutive cases. : Int J Periodontics Restorative Dent, v. 4, p. 381-89, 2003.

[5] Helmus, M. N., Tweden K.; "*Encyclopedic handbook of biomaterials and bioengineering, Part A*" - D. L. Wise, D. J. Trantolo, D. E. Altobelli, M. J. Yaszemski, J. D. Gresser, E. R. Schwartz - (NewYork: Marcel Dekker) – 1995.

[6] Moraes, P.C. Biocompatibilidade e comportamento do cimento de fosfato de cálcio, reforçado com diferentes tipos de fibras, implantado no rádio de coelhos.Tese (Mestrado)- Faculdade de Ciências Agrárias e Veterinárias, Universidade Estadual Paulista, Jaboticabal, SP. 73f. 2002.

[7] Kawachi,E.Y. et al.Biocerâmicas : tendências e perspectivas de uma área interdisciplinar. Química Nova, v.23, n. 4, p. 518-522, 2000.

[8] Campos L.E.C.*Análise físico-química e biológica de duas superfícies de implantes osteointegráveis e uso odontológico.* 138f. Dissertação (Mestrado) – Universidade Veiga de Almeida, Rio de Janeiro, 2007.

[9] Dorozhikin, S. V.; Epple, M. Biological and medical significance of calcium phosphates. Angewandte Chemie International Edition, v. 41, p. 3130-3146, 2002.

[10] Franco, K.L.; Borges, A.P.B.; Vilória, M.I.V. et al. Hidroxiapatita sintética pura, hidroxiapatita associada ao colágeno e hidroxiapatita sintética associada ao lipossoma como substitutos ósseos em defeitos provocados na tíbia de cães: aspectos da osteointegração à microscopia de luz transmitida. *Arq. Bras. Med. Vet. Zootec.*, v.53, p.431-436, 2001.

[11] Legeros, R.Z. Properties of osteoconductive biomaterials: calcium phosphates. *Clin. Orthop. Rel. Res.,* v.395, p.81-98, 2002.

[12] Borges, A.P.B. O efeito da hidroxiapatita sintética na regeneração óssea de defeito *provocado experimentalmente no terço proximal da tíbia de cães:* estudo clínico cirúrgico, radiológico e histológico por microscopia de luz e microscopia eletrônica de retrodisperçãoTese (Doutorado)-Escola de Veterinária, Universidade Federal de Minas Gerais, Belo Horizonte, MG. 112f. 1998.

[13] Ignácio, H.; Mazzer, N.; BarbierI, C. H.; Chierice, G. O. Utilização da poliuretana da mamona nas formas compacta e porosa no preenchimento de falha óssea: estudo experimental em cães. Rev. Bras. Ortop., v. 37, n. 5, p. 187-94, 2002.

[14] RumpeL,E.;Wolf,E.; Kauschke,E.; Bienengrber,V.; Bayerlein,T.; Gedrange,T.; Proff, P.; The biodegradation of hydroxyapatite boné graft substitutes in vivo. *Folia Morphologica,* v.65, n.1, p. 43-48, 2006.

[15] Ripamonti, U.; Duneas, N. Tissue Engineering of Bone by Osteoinductive Biomaterials. *MRS Bulletin*, v. 21, n. 11, Nov. 1996.

[16] Boyan, B.D. *et al.* Role of material surfaces in regulating bone and cartilafe cellresponse. *Biomaterials*, v. 17, p. 137-146, 1996.

Journal of Physics: Conference Series **313** (2011) 012006 doi:10.1088/1742-6596/313/1/012006

Growing axons analysis by using Granulometric Size Distribution

Mariela A. Gonzalez [1,2], Virginia L. Ballarin [1], Melina Rapacioli [2,3], A. R. Celín [2,3], V. Sánchez[2,4] y V. Flores [2,3,4]

[1] Measurement and Signal Processing Laboratory, School of Engineering, UNMdP, Juan B. Justo 4302, Mar del Plata, B7608FQD, ARGENTINA
[2] CONICET (Comisión Nacional de Investigaciones Científicas y Tecnológicas), Rivadavia 1917, Ciudad Autónoma de Buenos Aires, C1033AAJ, ARGENTINA.
[3] Interdisciplinary Group in Theoretical Biology, Departament de Bioestructural Sciences, UF. ARGENTINA.
[4] Institute of Cell Biology and Neurosciences "Prof. E. De Robertis", U.B.A. ARGENTINA.

mazulgonzalez@fi.mdp.edu.ar

Abstract. Neurite growth (neuritogenesis) in vitro is a common methodology in the field of developmental neurobiology. Morphological analyses of growing neurites are usually difficult because their thinness and low contrast usually prevent to observe clearly their shape, number, length and spatial orientation. This paper presents the use of the granulometric size distribution in order to automatically obtain information about the shape, size and spatial orientation of growing axons in tissue cultures. The results here presented show that the granulometric size distribution results in a very useful morphological tool since it allows the automatic detection of growing axons and the precise characterization of a relevant parameter indicative of the axonal growth spatial orientation such as the quantification of the angle of deviation of the growing direction. The developed algorithms automatically quantify this orientation by facilitating the analysis of these images, which is important given the large number of images that need to be processed for this type of study.

1. Introduction

Analysis of neurites outgrowth from explants of developing nervous tissue in culture media is a common experimental strategy to analyze the factors incluencing or regulating neuritogenesis (dendro- and axonogenesis), the directed growth cone migration, the directed neurite growth and the establishment of projection pathways and/or local neuronal circuits in the developing central nervous system [1][2]. Developing neurites are subtle dynamic structures. Their changing morphological characteristics make them difficult to measure and quantify. Thus, their relevant features, i.e. shape, thickness, length, spatial orientation etc, are usually described qualitatively. However, a relevant aspect in developmental neurobiology is the statistical analysis of neuritogenesis and this requires precise records of the above mentioned characteristics [3]. This situation reveals the need for designing and implementing algorithms with the ability to precisely reveal some these neurite features.

The main object of this work was the application of digital image processing algorithms to images obtained from cultures of explants obtained from a specific superior area of the chick central nervous system in order to automatically characterize the shape, size and spatial orientation of the growing axons.

Images obtained from this kind of tissue cultures are usually characterized by noisy background, non-homogeneous gray level (ilumination), groups of neuronal bodies, isolated migrating neurons and a large number of neurites with different characteristics depending on the embryonic stage and on the area under analysis. Axons usually differ in shape, gray level, texture, size, orientation etc. All these features greatly difficult the acquisition of a binary image [4].

Conventional image processing methods for analysis of shapes of objects requires a binary image and a subsequent calculation of factors roundness, shape, area, etc. [4][5]. Unlike standard techniques, morphological techniques are based on concepts of geometry, algebra, topology and set theory, to characterize structural properties in images [6]-[10]. The central idea of these techniques is to examine the geometric structures in an image overlaying them with small patterns called structuring elements. Of all mathematical morphology analysis techniques, the most appropriate tool for discriminating shapes is the granulometric size distribution. [11][13].

This paper proposes the computation of the granulometric size distribution with structuring elements of various shapes and sizes to quantify the number and orientation of the growing axons and and the subsequent statistical analysis.

2. Materials

To evaluate the results of the application of the granulometric size distribution, in a first step we used synthetics images, and in a second step we used images corresponding to different patterns of axons distribution commonly found in tissue cultures analyzed with the light microscope.

For the first step, we generated 30 synthetic images, 10 with a white background, 10 with light gray background and 10 with dark gray background. For each background, rectangles of different thicknesses with orientations of 0 °, 45 °, 90 ° and 135 °. The number of rectangles in each image was also variable. In Figure 1 we can see some examples of synthetic images.

In the second step we analysed 16 images of growing axons in tissue culture. Some of those images show clear preferential spatial orientations; others images show irregularly distributed spatial orientations. Images without axons were also analyzed. Figure 2 shows three cases representative of the growth of axons.

All algorithms were implemented using Matlab R14.

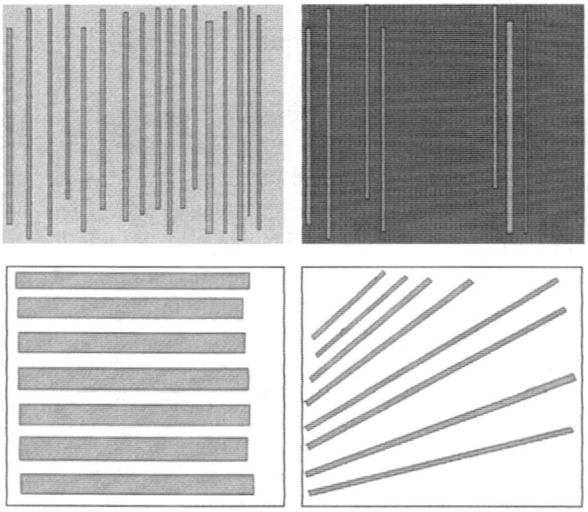

Figure 1. Synthetic Images

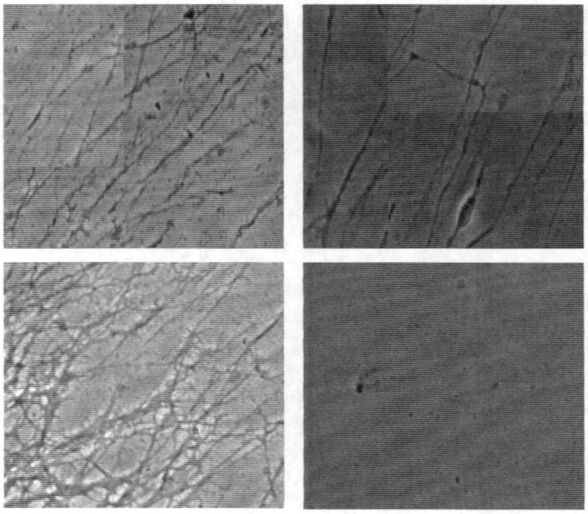

Figure 2. Growing axons Images

3. Methods

3.1. Mathematical Morphology applied to binary images

The basic operations of binary morphology are erosion and dilation [6]. These operations compare subsets of the binary image with a pattern element called structuring element. The structuring element is moved over the whole image [8]-[10].

Formally, the erosion of a set $A \subset \Re^2$ by a structuring element $E \subset \Re^2$ is defined as:

$$A \Theta E(x) = \{x \in A : E_x \subset A\} \tag{1}$$

which represents the translation of the set E at x. The result of erosion is a binary image where pixels are white if the structuring element is included within the subset of the analyzed image and black otherwise.

Similarly, dilation of a ser $A \subset \Re^2$ by a structuring element $E \subset \Re^2$ is defined as:

$$A \oplus E(x) = \{x \in A : E_x \cap A \neq \varnothing\} \tag{2}$$

The result of dilation is a binary image where pixels are white if at least exists one pixel of intersection between the subset of the original image and structuring element and black when the intersection is empty. Dilation is an operation of expansion while erosion is a contraction operation.

These basic operations define new morphological operators. For example, opening is defined as erosion followed by dilation:

$$A \circ E(x) = (A \Theta E) \oplus E \tag{3}$$

Closing is defined as dilation followed by erosion, that is:

$$A \bullet E(x) = (A \oplus E) \Theta E \tag{4}$$

From these operators, the granulometric size distribution is defined. This mathematical morphologic function is a tool for characterization of shapes and sizes of objects in an image. Given the family $\{A \circ E_\lambda\}_{\lambda \in I}$ obtain by the application of successive openings with structuring elements of increasing size is defined by the name of granulometric size distribution function is defined by:

$$G(\lambda) = 1 - \frac{\Omega(A \circ E_\lambda)}{\Omega(A)} \tag{5}$$

where Ω is a measure of the resulting image. For the case of binary images, Ω results to be the area.

3.2. Mathematical Morphology applied to gray level images

As in mathematical morphology for binary images, the basic morphological operations for gray level images are erosion and dilation [8]-[10].

Given two images f and g, with domains $D_f, D_g \subset \Re^2$, erosion of image f by structuring element g is defined as:

$$f \Theta g(s,t) = \min_{\substack{(x,y) \in Dg \\ (s+x,t+y) \in Df}} \{f(s+x, t+y) - g(x,y)\} \tag{5}$$

For each pixel of the image, the erosion operation is defined as the minimum difference between the intensities of the shifted structuring element and the corresponding intensities of the original image.

Similarly, the dilation of image f by structuring element g is defined as:

$$f \oplus g(s,t) = \max_{\substack{(x,y) \in Dg \\ (s-x,t-y) \in Df}} \{f(s-x, t-y) + g(x,y)\} \tag{6}$$

Dilation of each pixel of the image is defined as the maximum value of the sum between the intensities of the subset of the original image and the intensities corresponding structuring element.

The opening operator and the granulometric size distribution are defined similarly to the binary operators [14]-[15].

$$G(\lambda) = 1 - \frac{\Omega(A \circ E_\lambda)}{\Omega(A)} \tag{7}$$

Successive openings are made to the original image with increasing structuring elements. Then a measure Ω is calculated for each opening. In the case of gray level images, Ω is the volume. This result is normalized to the original image.

After calculating the granulometric size distribution function, which is actually a probability distribution function, mean, standard deviation and energy are calculated.

4. Results

The granulometric size distribution function was calculated for test images using structuring elements in different orientations. The variability of the moments was analysed in function of the background color, the number of rectangles, the width of the rectangles and their orientation.

A. Analysis of the variation of mean with respect to the variation of gray level of the background

In synthetic images with different gray levels for the same kind of rectangles (Fig.1), mean values of the granulometric size function were measured. This parameter showed a high variability without following any specific behaviour. Some of the results are shown in Table I.

TABLE I
MEAN VALUE OF THE GRANULOMETRIC SIZE DISTRIBUTION FUNCTION IN IMAGES WITH DIFFERENT BACKGROUNDS.

structuring element	white background 4 thin rectangles at 45°	light gray background 4 thin rectangles at 45°	dark gray background 4 thin rectangles at 45°
line 0°	0,3864	0,5416	0,9499
line 90°	0,6249	0,6233	0,8805
	white background 17 thin rectangles at 45°	light gray background 17 thin rectangles at 45°	dark gray background 17 thin rectangles at 45°
line 0°	0,8593	0,8175	0,8749
line 90°	0,4689	0,4345	0,3664
	white background 7 thick rectangles at 0°	light gray background 7 thick rectangles at 0°	dark gray background 7 thick rectangles at 0°
line 0°	0,5054	0,3987	0,5439
line 90°	0,891	0,8423	0,9139

B.Analysis of the variation of the mean with respect to the number and thickness of rectangles.

The mean value calculated from the size distribution function increases with the number and thickness of the rectangles in all test images analyzed. However, we cannot distinguish whether this increase is due to the increase in the number or in the thickness. Only this function could be used to assess the occupancy of axons in the image, ie the ratio between the area occupied by neurons and the background area. Table II shows these results.

TABLE II

MEAN VALUE OF THE GRANULOMETRIC SIZE DISTRIBUTION FUNCTION IN IMAGES WITH DIFFERENT NUMBER OF AXONS. (WITH A LINEAR STRUCTURING ELEMENT AT 0°)

white background 4 thin rectangles at 45°	white background 8 thin rectangles at 45°
0,5406	0,6536
light gray background 4 thin rectangles at 45°	light gray background 8 thin rectangles at 45°
0,5416	0,6521
dark gray background 4 thick rectangles at 135°	dark gray background 8 thick rectangles at 135°
0,8912	0,9021
white background 7 thin rectangles at 90°	white background 7 thin rectangles at 90°
0,6673	0,8593
light gray background 7 thin rectangles at 90°	light gray background 7 thin rectangles at 90°
0,6597	0,8175

C. Analysis of the variation of the mean with respect to the orientation of the rectangles and the orientation of the structuring element used

For the different synthetic images the granulometric size distribution was calculated with linear structuring elements with different orientations.

The results showed that when the structuring element used corresponds to the inclination of the rectangles of the images, the average value of the function is lower than in the other case. This is because when performing an opening with a structuring element which coincides with the shape of objects in the image, these objects do not appear in the resulting image, so that the measured volume is smaller and therefore the associated moment.

Table III shows some examples. This table highlights the values of the structuring element whose size distribution function coincides with the orientation of the rectangles. It is worth to note that this result was confirmed for the 30 test images.

TABLA III

MEAN VALUE OF THE GRANULOMETRIC SIZE DISTRIBUTION FUNCTION IN IMAGES WITH STRUCTURING ELEMENTS IN DIFFERENT DIRECTIONS.

inclination of the rectangles in the image	linear structuring element at 0°	linear structuring element at 90°	linear structuring element at 135°
90°	0,8175	0,4345	0,8319
90°	0,8153	0,6014	0,7284
135°	0,7671	0,5714	0,5497
90°	0,8749	0,3664	0,9057
90°	0,9078	0,4639	0,8846
135°	0,8912	0,8788	0,8075
90°	0,8924	0,3735	0,8657
90°	0,9064	0,4814	0,8813
90°	0,8593	0,4689	0,8789
90°	0,6673	0,4669	0,7698
0°	0,5054	0,891	0,8658
0°	0,3987	0,8423	0,8615

D.Analysis of the images of growing axons

Having analyzed the results for synthetic images, we evaluated the statistical on images of growing axons in tissue culture (Fig. 2). As in previous studies, the calculation of the granulometric size distribution function with different structuring elements permitted to know the orientation of the objects in the images. The mean value determined from this function is lower when the axon has the same orientation as the structuring element used in the calculation of the granulometric function.

The mean value calculated from the granulometric function also increases with the number and the thickness of the growing axons.

The granulometric function was calculated using a square structuring element. When there is no definite orientation of the axons, the average value calculated from the granulometric function is lower than in the other case. This feature allows the characterization of axonal networks where axons are irregularly distributed without any preferential orientation, since this structuring element analyzes all directions simultaneously.

TABLE IV

MEAN VALUE OF THE GRANULOMETRIC SIZE DISTRIBUTION FUNCTION IN IMAGES WITH GROWING AXONS.

(IMAGES WITHOUT AXONS AND WITH AXONS ALIGNED AT 0°, 45°, 90° AND 100°)

image / structuring element	axons at 45°, 90° and 100°	aligned axons at 45°	aligned axons at 0°	without axons
0°	0,7872	0,759	0,5559	0,6516
45°	0,7338	0,6956	0,6012	0,6425
90°	0,725	0,7387	0,5973	0,6368
135°	0,7325	0,7413	0,5967	0,6402
disk 3x3	0,6776	0,7804	0,6893	0,5007

Journal of Physics: Conference Series **313** (2011) 012006 doi:10.1088/1742-6596/313/1/012006

TABLE V
MEAN VALUE OF THE GRANULOMETRIC SIZE DISTRIBUTION FUNCTION IN IMAGES WITH GROWING AXONS (WITH AXONS ALIGNED AT 45° AND 90°)

image / structuring element	aligned axons at 45°	high number of aligned axons at 90°	low number of aligned axons at 45°	high number of aligned axons at 45°
0°	0,6519	0,7788	0,5777	0,7754
45°	0,6077	0,7256	0,5669	0,7327
90°	0,6509	0,6776	0,6187	0,7836
135°	0,6759	0,7155	0,5953	0,7965
Disk 3x3	0,6945	0,7125	0,5987	0,7913

Tables IV and V show the discrimination power of the mean value calculated from the granulometric function, in images of growing axons.

5. Conclusions

The present paper is a preliminary attempt to design a reliable method for digital image analysis with the ability to easily identify growing axons and to quantify relevant parameters of axonal growth indicative of their plasticity, dynamism and sensitivity to different factors during neuritogenesis (growth factors, extracellular matrix components, axon-neuron interactions, axon-glial cell interactions etc).

Our results show that the granulometric size distribution function is a useful tool for the characterization of the spatial orientation of objects both in synthetic images and in real images of growing axons in tissue culture.

The mean value obtained by means of this function allows a precise quantification of the angle of deviation of the axons longitudinal axes which constitute a reliable parameter indicating the preferential spatial orientation or the preferential growth direction.

This utility is not relevant when the growing axons fasciculate (parallel growing direction and close contact between adjacent axons) displaying a clearly defined preferential growth direction.

The granulometric size distribution function, however, can be advantageously applied (with great utility) in those cases when the growing axons do not display a preferential growth direction but they spatially arrange as irregular complex axonal networks. Even in these cases the granulometric function allows the detection and quantification of axons displaying different growth direction. This is a remarkable utility of the granulometric function since the most frequent images of axons growing in tissue culture correspond to complex networks of axons displaying none preferential spatial orientation.

The proposed automatic algorithms simplify the analysis of these images, which is important due to the large number of images that need to be processed for a robust study.

As future work we plan to implement the algorithms developed to characterize, in vivo, the degree of order or variability of growing axons with respect to the intrinsic spatial reference axis of the developing central nervous system (cephalic-caudal, dorsal-ventral and radial axes) as well as the time or stage-dependent changes in these parameters during the central nervous system development. These data could help to better understand the neuritogenic process and the development and stabilization of the neuronal circuits.

6. References

[1] S. Fiszer de Plazas, M. Rapacioli, D. Rodriguez, M. Vacotto, V. Flores, "Acute hypoxia differentially affects the gamma-aminobutyric acid type A receptor alpha(1), alpha(2), beta(2), and gamma(2) subunit mRNA levels in the developing chick optic tectum: Stage-dependent sensitivity". *Journal of Neuroscience Research,* vol. 95 No. 5 pp.280-286, 2008.

[2] R. Farías, C. Vidal, M. Rapacioli, V. Flores, "Basics elements for modelling the dynamics of cell migration in cell culture", *Journal of Physics: conference series.* vol. 90 pp.1-7, 2007.

[3] J. Mazzeo, M. Rapacioli, V. Sanchez, G. Scicolone, C. D'Attellis, V. Flores, "Characterizing cell proliferation process in the developing central nervous system", *InterJournal Complex Systems*, No. 1686 pp.1-7, 2006.

[4] Glasbey, C. A. y Horgan G. W., *Image analysis for the biological science, Statistics in Practice*, Series Editor Vic Barnett., Jonh Wiley and Sons, 1994.

[5] González, R y Woods, R., *Tratamiento Digital de imágenes*, Addison Wesley, 1996.

[6] Facon, J., Morfología *Matemática. Teoría y ejemplos*, Curitiba Brasil, CITS, 1996.

[7] Castleman K. R., Digital *Image Processing*, Prentice Hall, 1979.

[8] Serra, J. , *Image Analysis and Mathematical Morphology,* Academic Press, London, 1982.

[9] Serra, J., *Image Analysis and Mathematical Morphology- Part II: Theoretical Advances,*, Academic Press, London, 1988.

[10]Serra J. *Image analysis and Mathematical Morphology*, Academic Press, 1992.

[11]Ballarin, V. y Valentinuzzi, M.: "Segmentación en imágenes de Resonancia Magnética de Cerebro utilizando Morfología Matemática," Actas del Congreso Argentino de Bioingeniería. Tafí del Valle. Septiembre 2001. (Publicadas en CD), 2001.

[12]Heijmans, H., "'Theoretical aspects of gray-scale morphology," *IEEE Trans. Pattern Anal. Mach. Intell.,* vol. 13, pp.568-582, 1991.

[13]Vincent L., Dougherty E., *Morfological Segmentation for Textures and Particles*, Digital Image Processing Methods, Rochester, New York, 1994.

[14] Dougherty, E. R. y Astola J., *An introduction to nonlinear image prossesing*, Tutorial Texts in optical engineering, vol. TT16, 1994.

[15] Marshall, S. y Sicuranza G., *Advances in nonlinear signal and image prcessing*, Eurasip, 2006.

Journal of Physics: Conference Series **313** (2011) 012007 doi:10.1088/1742-6596/313/1/012007

Characterization of new materials for fiberoptic dosimetry

P Molina[1,2], M Santiago[1,2], J Marcassó[1,2], E Caselli[1,3], M Prokic[4], N Khaidukov[5] and C Furetta[6]

[1] IFAS, Facultad de Ciencias Exactas, Universidad Nacional del Centro de la Provincia de Buenos Aires, Pinto 399, 7000 Tandil, Argentina.
[2] Consejo Nacional de Investigaciones Científicas y Técnicas (CONICET), Rivadavia 1917, 1033 Buenos Aires, Argentina.
[3] Comisión de Investigaciones Científicas de la Provincia de Buenos Aires (CICPBA), calle 526 entre 10 y 11, 1900 La Plata, Argentina.
[4] Institute of Nuclear Sciences, Vinca, P.O. Box 522, 11000 Belgrade, Serbia.
[5] Kurnakov Institute of General and Inorganic Chemistry, RAS, 31 Leninskii Prospekt, 119991 Moscow, Russia.
[6] Touro University Rome, Division of Touro College New York, Circ.ne Gianicolense 15-17, 00153 Rome, Italy.

pmolina@exa.unicen.edu.ar

Abstract. In this work we have investigated the radioluminescence (RL) characteristics of three materials (Mg2SiO4:Tb, CsY2F7:Tb and KMgF3:Sm) in order to determine whether they can be used as real time dosimeters in the the framework the fiberoptic dosimetry (FOD) technique. This technique is based on the use of scintillating materials coupled to the end of an optical fiber, which collects the light emitted by the scintillator during irradiation. Since usually the intensity of the emitted light is proportional to the dose-rate, the technique provides a reliable measuring method, which can be employed in radiotherapy treatments.

1. Introduction

In the last years the implementation of new radiotherapy treatments techniques based on the use of high dose rates delivered in small treatment regions claim for the development of dosimetry systems suitable for in-vivo and real-time dose measurement having high spatial resolution [1]. The use of diodes and thermoluminescence dosimeters (TLD) constitutes a well established method for in-vivo dosimetry. Disadvantages of diodes such as energy or position dependence respect to the irradiation beam and the impossibility of using TLD in real time dose assessment makes it necessary to investigate alternative dosimetry techniques. For this reason dosimetry techniques based on the use of MOSFET [2]-[5], radiochromic films [6]-[8], diamond detectors [9][10], and fiberoptic dosimetry (FOD) [11][12] have been investigated.

The FOD technique takes advantage of the radioluminescence (RL) or scintillating phenomena observable in several dielectric or semiconductor materials when they irradiated with ionizing radiation (X-ray, gamma ray, electrons, etc…). The basic idea of the FOD technique is to attach a small (1mm^3) scintillating material to the extreme of an optical fiber placed in the region of interest where the dose rate is to be measured. The optical fiber transports the emitted light outside the irradiation bunker where a light detector measures its intensity. Since the intensity of the RL emission is generally proportional to the dose rate then it is possible to obtain a real-time, absolute measurement of the dose rate after calibration.

The main characteristics of the FOD technique are: a) high spatial resolution, due to the small size of the RL detector; b) is a passive detector, there is no need for electrical connections; c) it is a robust system suitable for clinical use; finally d) on-line acquisition, this means real time dose rate measurement [13].

Several materials have been investigated in order to be used as scintillators in FOD, but aluminum oxide doped with carbon (Al_2O_3:C) has shown the most promissory results so far because of both its high efficiency (RL intensity vs. dose rate) with respect to other materials [14]-[16] and commercial availability (provided by Landauer Inc.). One of the drawbacks of Al_2O_3:C is the accumulated dose dependence. To overcome this problem a computational algorithm for real-time recalibration the RL signal and periodical optical resets have been proposed [15].

The main drawback of the FOD technique is the spurious luminescence (commonly known as stem effect) that adds to the RL emission from the scintillator. The stem effect is the result of the intrinsic luminescence of the optical fiber and the luminescence from the Cherenkov radiation. The main component of the stem effect in PMMA (polymethy methacrilate) optical fibers used in FOD is the Cherenkov radiation. This radiation is generated in an optical fiber when relativistic charged particles pass through the core at speeds greater than the local speed of light. It shows a λ^{-3} wavelength dependence, being significant at the blue region of the visible spectrum [17]. The intensity of the Cherenkov radiation depends on the length and position of the irradiated portion of the optical fiber with respect to radiation beam. For this reason its contribution is not univocally related to the dose rate at the position of the scintillator. It is necessary to remove the contribution of the Cherenkov radiation in order to have a precise estimation of dose rate.

Several methods have been proposed to eliminate or at least reduce the Cherenkov radiation. The simplest one consists in employing a second optical fiber (dummy) having no scintillator at its extreme, which is placed next to the fiber having the scintillating material (detector). During irradiation, the dummy collects the light from Cherenkov radiation. This light is simultaneously subtracted to the RL emission that comes from the RL detector. The only inconvenient of this method is the increased size of the detector system, which affects the spatial resolution (two optical fibers instead of one). This issue is of importance in radiotherapy treatments where high dose gradients are expected. Another simple method to reduce the Cherenkov radiation is optical filtering: a filter is used to eliminate the wavelengths outside the interval of wavelengths emitted by the detector. Since the Cherenkov light decreases as λ^{-3} the best solution is to choose a scintillator emitting in the red-near infrared region. Thus by employing a high-pass filter most of the Cherenkov light is eliminated. The third method to reduce the Cherenkov contribution is known as pulsed method. This technique is useful when a pulsed irradiation source is used, such as a linear accelerator (LINAC). This method takes advantage of the fact that the Cherenkov radiation emission is only observable during irradiation pulses and that in some scintillators the RL emission is still measurable between irradiation pulses. For this reason, if the RL is acquired during the time interval between irradiation pulses the contribution of the Cherenkov radiation can be eliminated. This method have been successfully employed in FOD technique based on Al_2O_3:C detectors under LINACs irradiation [13], [15], [18], [19].

The main goal of this work has been to find scintillating materials suitable to be used in the FOD technique. In particular, the dependence of the RL intensity as function of the irradiation time and the RL spectra of different RL materials have been compared to similar measurements obtained with Al_2O_3:C. Emission spectra have been obtained in order to determine whether the costless simple filtering method could be employed in each case. Percentage depth dose (PDD) obtained with the different scintillators have been compared with the PDD obtained with a standard ionization chamber (used as reference measurement).

2. Materials

Three different FOD probes were built for this work. In particular, terbium-doped silicate (Mg_2SiO_4:Tb), terbium-doped fluoride (CsY_2F_7:Tb) and samarium-doped fluoride ($KMgF_3$:Sm) were employed as scintillators. These materials were synthesized following the techniques described in refs. [20]-[22] respectively.

In order to fabricate the FOD probes pieces of 1mm3 were cut and glued to one of the ends of a plastic core optical fiber (PMMA, 980 microns diameter core, and 2 mm diameter outer jacket). The scintillators were optically shielded with an opaque, water resistant coating in order to avoid external light and humidity. A FOD probe was also fabricated using commercial Al_2O_3:C rods (1 mm diameter, 2 mm length, by Landauer Inc.). A fiber having no scintillator at its end was employed as a dummy.

Irradiation of the RL probes was performed in-situ (radiotherapy facility) employing a Theratron 80 60Co source, which renders 0.35 Gy min^{-1} at 5 mm water depth (80 cm SSD, source to surface distance) and an irradiation field of 10×10 cm^2. The RL signal from the FOD probes was measured by means of a Hammamatsu H9319 photon counting photomultiplier tube (PMT) having a sensitivity range between 300 and 850 nm.

All measurements were carried out in a water phantom (Civco MT-100) at room temperature under the following reference conditions, namely, 10×10 cm^2 field and 80 cm SSD. In all cases the end of the RL probe containing the scintillator (sensitive end) was placed at 5mm water depth and the fiber was perpendicularly oriented with respect to the beam axis.

The percentage depth dose (PDD) curve was determined at different depths in the water phantom by means of a manual depth dose apparatus (Civco, 0.1 mm resolution). PDD profiles were determined from surface down to 100 mm, under reference conditions. A PTW 30013 Farmer-type ionization chamber and a PTW UNIDOS E electrometer were used for reference dose-rate measurements. The stem effect contribution from the RL probes based on Mg_2SiO_4:Tb and CsY_2F_7:Tb scintillators were removed by subtracting the RL signal from the dummy probe measured in identical conditions. For the RL probe based on $KMgF_3$:Sm scintillators a high-pass Schott filter with a cutoff wavelength of 645 nm was used for stem effect removal.

The spectrum of the RL emission was measured by means of an Acton Research SP-2155 0.15 m monochromator with a resolution of 10 nm. The Hamamatsu H9319 photon counting placed at the exit slit was employed to detect the scattered light.

3. Results and discussions

Figure 1 shows the RL curves as function of time corresponding to the materials investigated in this work. It must be noted that a factor of 10 has been applied to both the RL curves of $KMgF_3$:Sm and the curve corresponding to the dummy. Although the samples have different transparency a reasonable comparison is possible since all the samples have the same volume (1 mm^3). As can be seen in figure 1 the rise time of the RL signal of the investigated probes reaches a maximum value very rapidly (tens of seconds). On the other hand, it takes almost a minute the Al_2O_3:C to reach its maximum intensity. It is worth mentioning that the RL signal from the dummy probe is always one order of magnitude lower than the RL signal from the different probes. The only exception is the $KMgF_3$:Sm probe, which shows a RL signal as intense as that of the dummy (stem effect). In all cases, the importance of the stem effect emission is enough to justify the use of removal techniques and the search for new scintillators materials having even higher RL efficiency.

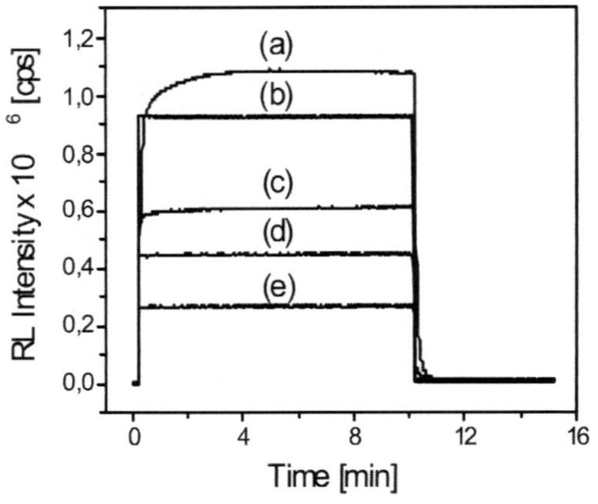

Figure 1. RL curves from Al_2O_3:C (a), CsY_2F_7:Tb (b), Mg_2SiO_4:Tb (c), $KMgF_3$:Sm × 10 (d) and dummy × 10 (e).

Figure 2 shows the emission spectrum of the different probes. The first spectrum (dummy) shows a maximum value near 400 nm and slowly decreases at higher wavelengths. The two next spectrums corresponding to the RL probes based on Al_2O_3:C and Mg_2SiO_4:Tb scintillators show an important overlap with the spectrum corresponding to the stem effect. This fact in precludes in these cases the use of the simple optical filtering technique for stem effect removal. However, these RL probes are in principle suitable to be used in LINACs employing the pulsed method to avoid the stem effect. This is not the case of the RL probe based on $KMgF_3$:Sm (figure 2(e)), whose emission spectrum shows a maximum value at 685 nm. By using a high-pass filter having a cutoff wavelength of 645 nm it is possible to reduce almost the entire contribution of the stem effect preserving the RL signal from the scintillator. However, its low RL efficiency makes this material not suitable for the FOD technique. The emission spectrum from the RL probe based on CsY_2F_7:Tb (figure 2(d)) shows a RL spectrum that makes possible the use of the simple optical removal technique in order to improve the RL signal/stem by employing, for instance, a long-pass filter with cutoff wavelength at 550 nm.

Figure 3 shows the comparison among the PDD curve obtained with the ionization chamber (solid line) and those from the different RL probes (scatter curves). All the PDD curves shows a maximum value at 5 mm water depth, as expected, except the RL probe based on CsY_2F_7:Tb that shows a maximum value at 16 mm water depth. It can also be seen in figure 3 that the PDD from the RL probe based on Mg_2SiO_4:Tb is similar to that corresponding to the ionization chamber. The PDD from the Al_2O_3:C RL probe shows a sub-response compared to the PDD curve from the ionization chamber. This result is consistent with the fact that the RL signal from this material varies as a function of accumulated dose during the successive measurements performed to obtain the PDD curve [15], [19].

The PDD from the $KMgF_3$:Sm-based RL probe shows a supra-response compared to the PDD curve from the ionization chamber. Investigations are in progress in order to determine whether the supra-response of this material is a dose-related effect. On the other hand, the PDD curve obtained with the RL probe based on CsY_2F_7:Tb shows an anomalous behavior compared to that obtained from the ionization chamber. The maximum value is shifted toward higher depths (16 mm) and the slope diverges from that of the PDD curve obtained with the ionization chamber. The origin of this unexpected behavior could be related to the particular physical process. Further basic research must be carried out in order to explain this scintillating phenomenon.

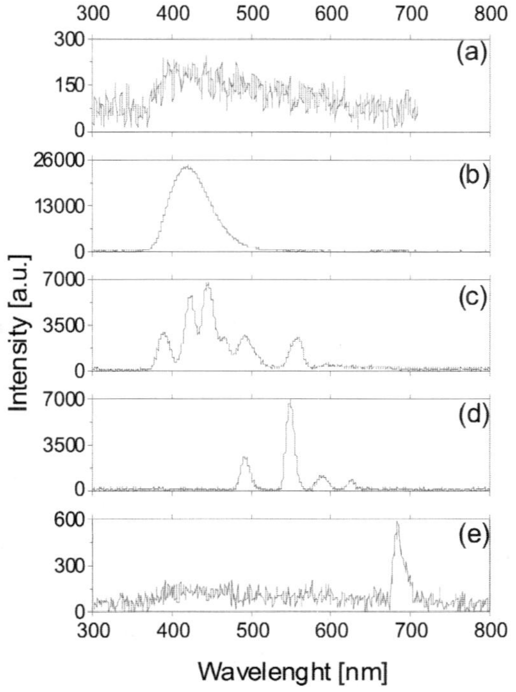

Figure 2. RL emission spectrum from (a) dummy (efecto stem); (b) Al_2O_3:C; (c) Mg_2SiO_4:Tb; (d) CsY_2F_7:Tb; and (e) $KMgF_3$:Sm.

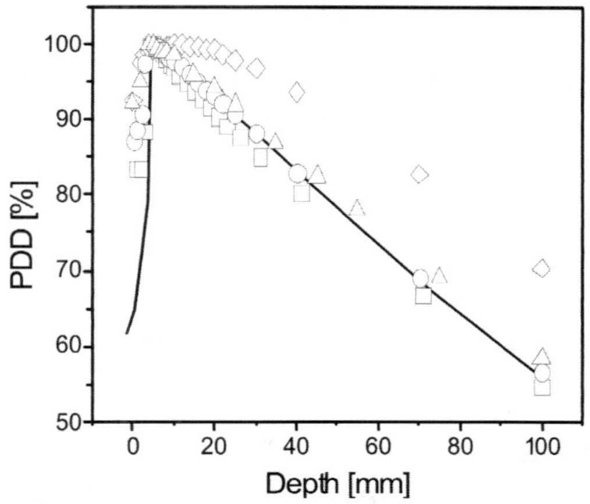

Figure 3. Percentage Depth Dose in water (PDD) from Al_2O_3:C (□), Mg_2SiO_4:Tb (○), CsY_2F_7:Tb (◊) and $KMgF_3$:Sm (Δ) compared to values obtain with a standard ionization chamber (solid line).

4. Conclusions

The optical fiber dosimetry is a promissory alternative technique for in-vivo dose assessment in real time. By means of this technique dose rate with high spatial resolution (lower than 1 mm) can be measured, as required by most recent radiation treatments. The FOD technique is mechanically robust and therefore suitable for routine handling by technicians in radiation treatment facilities.

In this work three materials have been investigated as possible scintillator to be employed in the context of the FOD technique: Mg_2SiO_4:Tb, CsY_2F_7:Tb and $KMgF_3$:Sm.

Terbium-doped silicate (Mg_2SiO_4:Tb) shows a high RL efficiency similar to that of the commercial compound Al_2O_3:C. Besides, the response of the Mg_2SiO_4:Tb shows no dependence on the accumulated dose. However, its application is limited to dosimetry in LINACs, where the pulsed method for stem effect removal is possible.

Samarium-doped fluoride ($KMgF_3$:Sm) has a poor RL efficiency, with an too low RL response/stem effect ratio.

Finally, terbium-doped fluoride (CsY_2F_7:Tb) has an anomalous RL response behaviour and it does not reproduce the absorption of energy in water, so this fluoride is not suitable for dosimetry in radiotherapy.

5. References

[1] Beddar A 2007 Plastic scintillation dosimetry and its application to radiotherapy *Radiat. Meas.* **41** S124–33

[2] Halvorsen H 2005 Dosimetric evaluation of a new design MOSFET in vivo dosimeter *Med. Phys.* **32** 110-17

[3] Lavallée M, Gingras L and Beaulieu L 2006 Energy and integrated dose dependence of

MOSFET dosimeter sesitivity for irradiation energies between 30 kV and 60Co *Med. Phys.* **33** 3683-89

[4] Lipovetzky J, Redin E and Faigón A 2007 Electrically erasable Metal Oxide-Semiconductor dosimeter *IEEE Trans. Nucl. Sci.* **54** 1244-50

[5] Faigon A, Bonafonte I, Esquerra A, Febrer and Vallverdu F 1997 A bilingual text-to-speech system in Spanish and Catalan *Proc. of the 5th European Conference on Speech Communication and Technology* **5** 2455-58

[6] Cheung K, Buston M and Yu P 2002 Multilayer Gafchromic film detectors for breast skin dose determination in-vivo *Phys. Med. Biol.* **47** N31-7

[7] Dempsey J, Low D, Mutic S, Markman J, Kirov A, Nussbaum G and Williamson J 2000 Validation of a radiochromatic film dosimetry system for quantitative two-dimensional imaging of acute exposure dose distribution *Med. Phys.* **27** 2462-75

[8] Yamauchi M, Tominaga T, Nakamura O, Veda R and Hoshi M 2004 Gafchromic film dosimetry with a flatbed color scanner for Leksell Gamma Knife therapy *Med. Phys.* **31** 1243-8

[9] Heydarian M, Hoban P, Beckham W, Borchardt I and Beddoe A 1993 Evaluation of a PTW diamond detector for electron beam measurements *Phys. Med. Biol.* **38** 1035-42

[10] Bucciolini M, Buonamici B, Mazzocchi S, De Angelis C, Onori S and Cirrone G 2003 Diamond detector versus silicon diode and ion chamber in photons beams of different energy and field size *Med. Phys.* **3** 2149-54

[11] Huston A, Justus B, Falkenstein P, Miller R, Ning H and Altemus R 2001 Remote optical fibre dosimetry *Nuc. Instrum. Methods Phys. Res. B* **184** 55-67

[12] Polf J, Yukihara E, Akselrod M and McKeever S 2004 Real-time luminescence from Al_2O_3:C fiber dosimeters *Radiat. Meas.* **38** 227–40

[13] Justus B, Falkenstein P and Huston A 2004 Gated fiber-optic-coupled detector for in vivo real-time radiation dosimetry *Appl. Opt.* **43** 1663

[14] Aznar M, Hemdal B and Medin J 2005 In vivo absorbed dose measurements in mammography using a new real-time luminescence technique *Br. J. Radiol.* **78** 328–34

[15] Damkjaer S, Andersen C and Aznar M 2008 Improved real-time dosimetry using the radioluminescence signal from Al_2O_3:C *Radiat. Meas.* **43** 893-7

[16] Marckmann C, Aznar M and Andersen C 2006 Influence of the stem effect on radioluminescence signals from optical fibre Al_2O_3:C dosimeters *Radiat. Prot. Dosimetry* **119** 363–7

[17] De Boer S, Beddar A and Rawlinson J 1993 Optical filtering and spectral measurements of radiation-induced light in plastic scintillation dosimetry *Phys.Med. Biol.* **38** 945-58

[18] Clift M, Johnston P and Webb D 2002 A temporal method of avoiding the Cherenkov radiation generated in organic scintillator dosimeters by pulsed mega-voltage electron and photon beams *Phys. Med. Biol.* **47** 1421–33

[19] Andersen C, Marckmann C, Aznar M, Botter-Jensen L, Kjaer-Kristoffersen F and Medin J 2006 An algorithm for real-time dosimetry in intensitymodulated radiation therapy using the radioluminescence signal from Al_2O_3:C *Radiat. Prot. Dosimetry* **120** 7-13

[20] Mittani J, Prokic M and Yukihara E 2008 Optically stimulated luminescence and thermoluminescence of terbium-activated silicates and aluminates *Radiat. Meas.* **43** 323–6

[21] Marcazzó J, Henniger J and Khaidukov N 2007 Efficient crystal radiation detectors based on Tb^{3+}-doped fluorides for radioluminescence dosimetry *J. Phys. D: Appl. Phys.* **40** 5055-60

[22] Patterson D 1962 Controlled Atmosphere Kyropoulos Growth of Alkali Halide Single Crystals *Review of Scientific Instruments* **33** 831-3

Journal of Physics: Conference Series **313** (2011) 012008 doi:10.1088/1742-6596/313/1/012008

Microstructural Assessment of Cancellous Bone Using 3D Microtomography

Silva A M H[1], Alves J M[1], Da Silva O L[1], Silva Junior N F[1], Gazziro M[1], Pereira J C[1], Lasso P R O[2], Vaz C M P[2], Pereira, C A M[3], Leiva T P[3], Guarniero R[3]

[1] Bioengineering Graduate Program, University of São Paulo, Sao Carlos, SP, Brazil

[2] EMBRAPA - Agricultural Instrumentation, São Carlos, SP, Brazil

[3] Institute of Orthopaedics and Traumatology, General Hospital, College of Medicine, University of São Paulo, SP, Brazil

alhakme@sc.usp.br, jma@sc.usp.br, orivaldo@sc.usp.br, nelsonjr@sc.usp.br, gazziro@sc.usp.br, lasso@cnpedia.embrapa.br, vaz@cnpdia.embrapa.br, robertoguarniero@clinicareboucas.com.br, lim.iot@hcnet.usp.br tomazpuga@hcnet.usp.br

Abstract. Cancellous bones have a porous microstructure and can be modeled as linear elastic solid, heterogeneous and anisotropic. Few studies regarding the morphometric analysis of trabecular bone samples with 3D microtomography have been published so far. The technique has spread worldwide for the characterization of trabecular structures in studies related to bone quality and its relationship with metabolic diseases bone like osteoporosis. In our study cancellous bone samples with cubic and cylindrical geometry were extracted from bovine femur were used to investigate the structural arrangement of bone through high resolution x-ray 3D microtomography (μCT). Four trabecular microstructural parameters (tissue volume, bone volume, bone volume fraction and tissue surface) were measured by 2D (stereological method) and 3D morphometric analysis using the software CTan Analyser supplied by the manufacturer of the microtomograph (SkyScan, model 1172, Belgium). The measurements were done in three main directions (superior-inferior, medial-lateral and anterior-posterior) to investigate the correlation between the 2D and 3D morphometric analysis. The results show a high correlation between the analysis. The x-ray 3D microtomography technique has a great potential for the assessment of bone quality.

1. Introduction

The assessment of macroscopic properties of biological tissues such as bone as well as the assessment of its microscopic structure is of great interest in the characterization of bone quality. The architecture of bone tissue varies with the region of the skeleton, which may be spongy, named cancellous or trabecular bone, or compact, named cortical bone. Cancellous bone has a porous microstructure with numerous small interconnected trabeculaes (100-300 mm of thickness, 300-1500 μm spacing between adjacent trabeculaes) [1] and a high surface area. It is predominantly found at the ends of long bones protected by a layer of cortical bone. It is also found in the internal structure of the calcaneus, the skull, pelvis and vertebrae of the spine. Approximately 20% of the skeleton is cancellous. The trabeculae tend to be oriented along the direction of principal stress in adaptation to external loading of the environment. Between 75-95% of the cancellous bone consists of interconnected pores filled with bone marrow. It has the ability of getting deformation prior to fracture when subjected to a load. Trabeculaes orientation in a sample of cancellous bone varies with direction, as shown in Figure 1. In these images the zy plane is a preferred orientation of the trabeculae and no preferred orientation can be seen in the planes xy and zx.

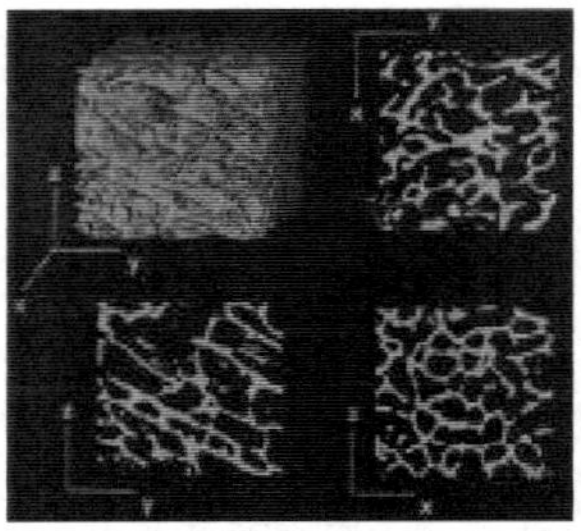

Figure 1: 3D images and cross sections of a cancellous bone sample obtained by 3D microtomography [2].

Trabecular bone density was related to the strength and stiffness [3-7]. Despite the density measurements often lead to reasonable estimates of mechanical properties, they do not explain certain variations in these properties nor are they influenced by the microstructure of cancellous bone, thus representing a scalar measure unable to totally predict the mechanical properties. Two bone samples with volume fractions (BV / TV = Bone Tissue Volume / Volume) close together, 18% and 19%, respectively, had mechanical properties that ranged from 30% to an order of magnitude, depending on the direction analyzed [5]. X-ray 3D microtomography (μCT) is a non-destructive technique that allows visualization of cross sections (internal cuts or slices) of an object [8], three dimensional reconstruction of the object as well as 2D and 3D morphometric analysis [9, 10]. The μCT technique has made significant contributions to the investigations on bone quality assessment [9-12].

2. Materials and methods

The Eighteen cylindrical samples of cancellous bone (10 mm of diameter and 20 mm of height) were obtained from six bovine femurs using the following methodology carried out at the Biomechanics Laboratory, Institute of Orthopaedics and Traumatology, General Hospital, College of Medicine, University of São Paulo at São Paulo and at the Mechanical Shop of the School of Engineering of Sao Carlos, University of São Paulo at São Carlos:

a) Freezing the femurs at -20 °C.
b) Searching at room temperature the main directions of the trabeculae using an x-ray radiograph of each femur.
c) Removal of a cubic trabecular sample of each femur with the trabeculae oriented in the main directions. The sample dimensions were approximately 30 mm x 30 mm x 30 mm. The removal used an apparatus that allowed angular and longitudinal movement of the femur (Fig. 2) and a circular saw (177.8 mm of radius and 3.2 mm of thickness) attached to a cutting machine working at 5500 rpm.

Figure 2: Apparatus for removal of the cubic sample.

d) Extraction of three cylindrical trabecular samples (10 mm of diameter and 20 mm of height) from each cube monitoring the temperature of the sample with a infrared digital temperature sensor (MT-350 Brand Minipa, output power below 1 mW, ± 0,5 ºC precision, 630-670 nm wavelength) to avoid necrosis of the bone tissue during the drilling process [13]. The extraction is illustrated in Fig. 3. The cylinder in each main direction was taken using a lathe with a radial drill working at low rotation where was attached a carbide cannula trephine with 10 mm internal diameter, 12 mm external diameter and 25 mm of height (Fig. 4).

e) Immersion of the cylindrical samples in saline solution (0.9% NaCl) until the structural quantification by 3D microtomography using the CT-Analyzer software supplied by mananu.

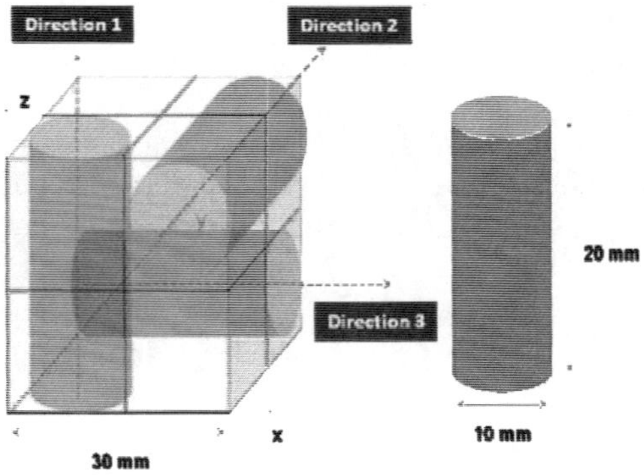

Figure 3: Schematic of the extraction of three cylinders in the three main directions.

Figure 4: Cannula trephine to extract the 10 mm x 20 mm cylinder.

The high resolution microtomograph (model 1172, SkyScan, Belgium) at EMBRAPA - Agricultural Instrumentation has a micro focus x-ray source operating at high voltage (100kV), a specimen holder with a x, y and z precision movement, and a 10MP (4000 x 2300 pixels) CCD camera detector. A computer host controls the data acquisition. A cluster of computers carry out the tomographic reconstruction. This system can be represented by the diagram in figure 5.

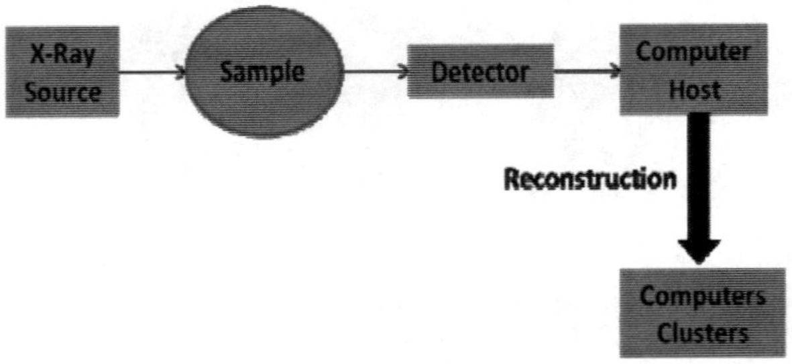

Figure 5: Microtomograph diagram [8]

The microstructural analysis of the cylindrical samples was held at EMBRAPA - Agricultural Instrumentation (Sao Carlos, SP, Brazil) with the following procedure:

a) Each cylindrical sample was kept in a circular template using "modeling clay" to get a vertical alignment, to avoid undesirable movements and maintain the sample in the visual field of the radiation detector (CCD camera).

b) The reconstruction of the entire volume of the samples (10mm x 20mm) couldn't be carried out due to problems with the computer memory. A volume of interest (VOI) with 1.4 mm of height (201 cross sections) from the central region of the sample was taken for the 2D and 3D morphometric analysis. The cross sections were were TIF/16 bits format images with a resolution of 6.7 μm for each pixel. The software CTan Analyzer version 1.7.0.0 was used in the analysis.

3. Results

Table 1 shows the cubic samples dimensions and the temperature (maximum and minimum) measured during the drilling procedure for getting the cylindrical samples.

Table 1 - Values of the parameters (cubic samples)

Cubic Sample	Dimension	Cylindrical Sample	Temperature (° C)	
			Minimum	Maximum
(Cube 1) 1st Pair right femur	23 x 26 x 25	1	18,0 ± 0,5	26,0 ± 0,5
		2	21,0 ± 0,5	33,0 ± 0,5
		3	20,0 ± 0,5	26,0 ± 0,5
(Cube 2) 1st Pair left femur	25 x 26 x 26	1	19,0 ± 0,5	42,0 ± 0,5
		2	20,0 ± 0,5	28,0 ± 0,5
		3	20,5 ± 0,5	24,0 ± 0,5
(Cube 3) 2nd Pair right femur	21 x 21 x 23	1	18,0 ± 0,5	27,0 ± 0,5
		2	18,0 ± 0,5	23,0 ± 0,5
		3	19,0 ± 0,5	21,0 ± 0,5
(Cube 4) 2nd Pair left femur	24 x 26 x 27	1	18,5 ± 0,5	26,0 ± 0,5
		2	19,0 ± 0,5	32,5 ± 0,5
		3	18,0 ± 0,5	23,0 ± 0,5
(Cube 5) 3rd Pair right femur	25 x 25 x 23	1	19,0 ± 0,5	25,0 ± 0,5
		2	20,5 ± 0,5	27,0 ± 0,5
		3	20,0 ± 0,5	23,0 ± 0,5
(Cube 6) 3rd Pair left femur	23 x 25 x 22	1	20,0 ± 0,5	26,0 ± 0,5
		2	19,0 ± 0,5	21,0 ± 0,5
		3	19,5 ± 0,5	21,5 ± 0,5

Among the eighteen samples nine were damaged in the drilling process. Two samples (Fig. 6 and 7), which couldn't be scanned in the same region of microtomography analysis were discarded.

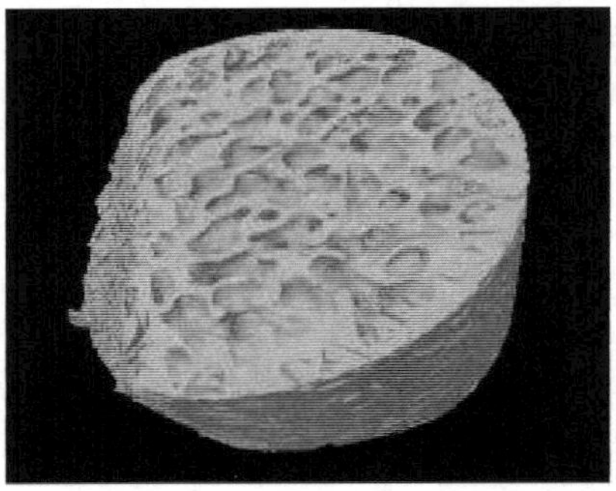

Figure 6: Sample 4, 3D reconstruction.

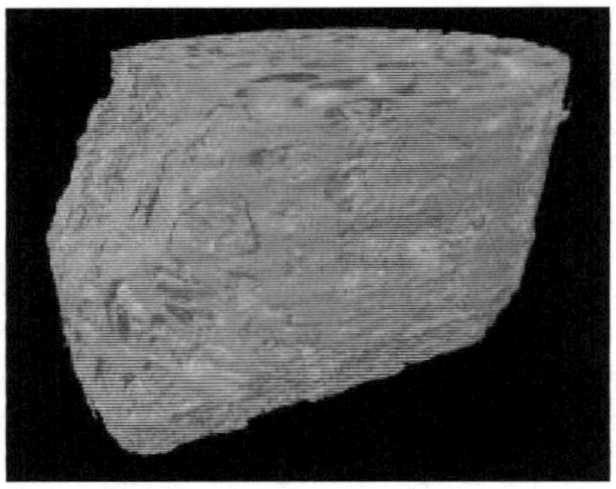

Figure 7: Sample 11, 3D reconstruction.

Tables 2, 3 and 4 show the averaged values for each structural parameter from the 2D and 3D morphometric analysis in the superior-inferior, lateral-medial and anterior-posterior main direction.

Table 2 - Average values of the microstructural paremeters from the 2D and 3D morphometric analysis in the superior-inferior direction (n = 6).

Parameters	Abbreviation	Unit	2D	3D
Tissue Volume	TV	mm^3	111,67	111,94
Bone Volume	BV	mm^3	54,69	54,39
Bone Volume Fraction	BV/TV	%	49,05	48,64
Tissue Surface	TS	mm^2	212,23	212,59

Table 3 - Average values of the microstructural paremeters from the 2D and 3D morphometric analysis in the lateral - medial direction (n = 5).

Parameters	Abbreviation	Unit	2D	3D
Tissue Volume	TV	mm^3	108,37	108,92
Bone Volume	BV	mm^3	51,10	51,17
Bone Volume Fraction	BV/TV	%	47,15	46,98
Tissue Surface	TS	mm^2	207,88	208,70

Table 4 - Average values of the microstructural paremeters from the 2D and 3D morphometric analysis in the anterior-posterior direction (n = 5).

Parameters	Abbreviation	Unit	2D	3D
Tissue Volume	TV	mm^3	110,00	110,42
Bone Volume	BV	mm^3	57,90	58,18
Bone Volume Fraction	BV/TV	%	52,53	52,68
Tissue Surface	TS	mm^2	210,82	211,26

The correlation between the structural values from the 2D and 3D morphometric analysis can be seen in figures 8 to 11. The correlation was calculated by the Statistica Software (version 8.0).

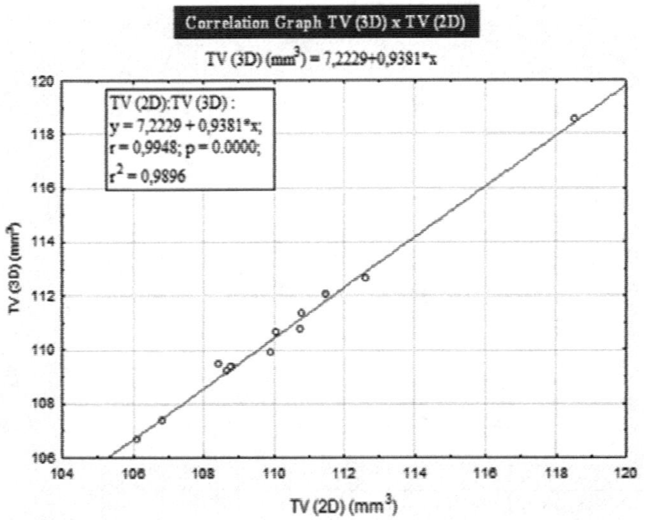

Figure 8: Correlation between bone tissue volume (TV) values from the 2D and 3D morphometric analysis.

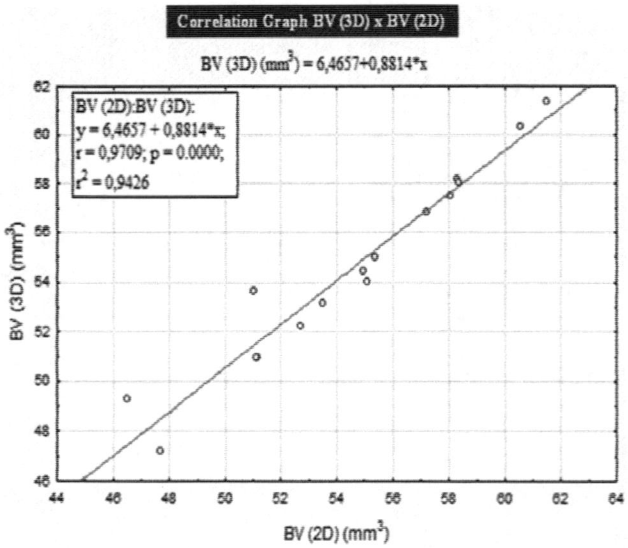

Figure 9: Correlation between bone volume (BV) values from the 2D and 3D morphometric analysis.

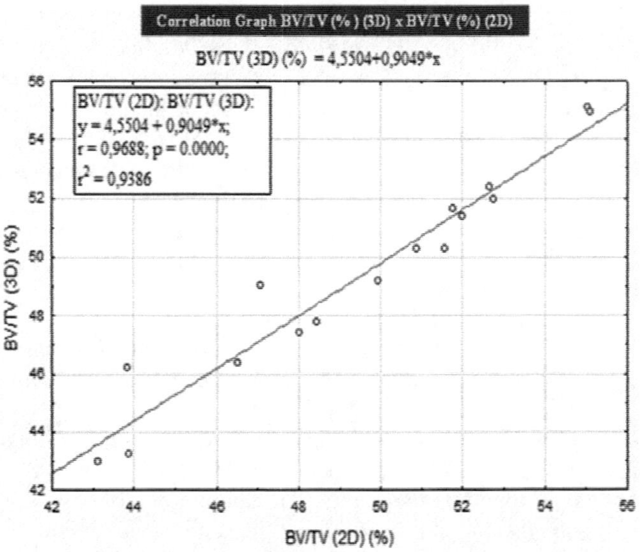

Figure 10: Correlation between bone volume fraction values (BV/TV) from the 2D and 3D morphometric analysis.

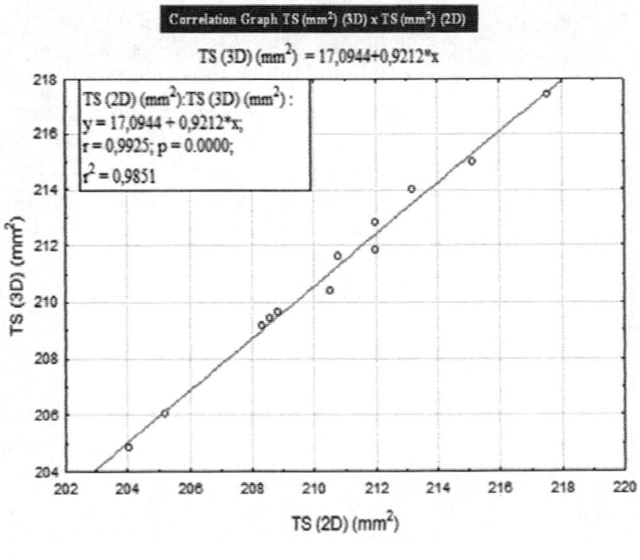

Figure 11: Correlation between bone tissue surface (TS) values from the 2D and 3D morphometric analysis.

4. Conclusion

There is a strong corrrelation between measurements of the structural parameters tissue volume (TV), bone volume (BV), bone volume fraction (BV/TV) and tissue surface (TS) measured by 2D and 3D morphometric analysis using x-ray 3D microtomography (μCT) and the CTan Analyser software supplied by the manufacturer of the microtomograph (SkyScan, model 1172, Belgium). The uCT is an important tool for the quantification of cancellous bone arquitecture.

Acknowledgments

This work was funded by CAPES (Brazilian Federal Research Foundation) and FAPESP (São Paulo Research Foundation, grant # 2008/56563-5). The authors would like to thank Paulo Henrique Grassano Murta and Nilo Narcizo de Souza for the donation of bovine femurs.

References

[1] Rodrigues MES, 2003, *Análise do risco de fratura óssea por ultrasonometria e ensaio mecânico de compressão* (Dissertação Mestrado) Escola de Engenharia de São Carlos/Faculdade de Medicina de Ribeirão Preto/ Instituto de Química de São Carlos (Universidade de São Paulo).

[2] Goulet RW, Goldstein SA, Ciabelli MJ, Kuhn JL, Brown MB, Feldkamp LA, 1994, *The relationship between the structural and orthogonal compressive properties of trabecular bone* (Journal of Biomechanics) vol 27 pp 375-389.

[3] Currey JD, 1969, *The mechanical consequences of variation in the mineral content of bone* (Journal of Biomechanics), vol 2 ,pp 1 – 11.

[4] Carter DR and Hayes WC, 1977, *The compressive behavior of bone as a two-phase porous structure* (Journal of Bone and Joint Surgery), vol 59, pp 954 – 962.

[5] Golsdstein SA, Wilson DL, Sonstegard DA, Matthews L, 1983, *The Mechanical properties of human trabecular bone as a function of metaphyseal location* (Journal of Biomechanics) vol 16 pp 959 – 969.

[6] Rice JC, Cowin SC, Bowman JA (1988), *On the dependence of elasticity and strength of cancellous bone on apparent density* (Journal of Biomechanics), vol 21, pp 155 – 168.

[7] Ciarelli MJ, Goldstein SA, Kuhn J L, Cody DD, Brown MB, 1991, *Evaluation of orthogonal mechanical properties and density of human trabecular bone from the major methaphyseal regions with materials testing and computed tomography* (Journal of Orthopaedic Research) vol 9 pp 674 – 682.

[8] Lasso PRO, Vaz CMP, Naime JM, 2008, *Recomendações para utilização do microtomógrafo de raios-X Sky Scan 1172 em ciências agrárias e biológicas* (Embrapa Instrumentação Agropecuária. Documentos, 37, p 22.

[9] Hildebrand T, Laib A, Muller R, Dequeker J, Ruegsegger P, 1999, *Direct Three-Dimensional Morphometric Analysis of Human Cancellous Bone: Microstructural Data from Spine, Femur Iliac Crest and Calcaneus* (Journal of Bone and Mineral Research) vol 14 (7), pp 1167-1174.

[10] Feldkamp LA, Goldstein SA, Parfitt M, Jesion G, Kleerekoper M, 1989, *The direct estimation of the three-dimensional bone arquitecture in vitro by computed tomography* (Journal of Bone and Mineral Research), vol 4, pp 3-11.

[11] Odgaard A, 2001, *Quantification of cancellous bone architecture* (Cowin S C, Bone Mechanics Handbook Informa HealthCare).

[12] Dreossi D and Cosmi E, 2007, *Numerical and experimental structural analysis of trabecular architecture* (Meccanica), vol 42, pp 85-93.

[13] Eriksson AR and Albrektsson T, 1984, *The effect of heat on bone regeneration: an experimental study in the rabbit using the bone growth chamber* (J Oral Maxillofac Surg.), vol 42 (11), pp 705-711.

Temporal pattern of pulse wave velocity during brachial hyperemia reactivity

S Graf[1,2], M J Valero[1], D Craiem[1,2], J Torrado[3], I Farro[3], Y Zócalo[3], G Valls[3], D Bía[3], R L Armentano[1]

[1] Facultad de Ingeniería y Ciencias Exactas y Naturales, Universidad Favaloro, Belgrano 1723, Buenos Aires C1093AAF, Argentina
[2] CONICET, Rivadavia 1917, Buenos Aires C1033AAJ, Argentina
[3] Departamento de Fisiología, Facultad de Medicina, Universidad de la República, General Flores 2125, Montevideo 11800, Uruguay

graf@favaloro.edu.ar

Abstract. Endothelial function can be assessed non-invasively with ultrasound, analyzing the change of brachial diameter in response to transient forearm ischemia. We propose a new technique based in the same principle, but analyzing a continuous recording of carotid-radial pulse wave velocity (PWV) instead of diameter. PWV was measured on 10 healthy subjects of 22±2 years before and after 5 minutes forearm occlusion. After 59 ± 31 seconds of cuff release PWV decreased 21 ± 9% compared to baseline, reestablishing the same after 533 ± 65 seconds. There were no significant changes observed in blood pressure. When repeating the study one hour later in 5 subjects, we obtained a coefficient of repeatability of 4.8%. In conclusion, through analysis of beat to beat carotid-radial PWV it was possible to characterize the temporal profiles and analyze the acute changes in response to a reactive hyperemia. The results show that the technique has a high sensitivity and repeatability.

1. Introduction

The vascular endothelium is a large paracrine organ that secrets numerous factors regulating vascular tone, cell growth, leukocyte interactions and thrombogenicity. The endothelium senses and responds to a myriad of internal and external stimuli through complex cell membrane receptors and signal transduction mechanisms, leading to synthesis and release of various vasoactive, thromboregulatory and growth factor substances. Endothelial function is altered by many pathophysiological insults, such as tobacco, arterial hypertension and diabetes. It is known that the endothelial dysfunction play a seminal role in atherogenesis [1], [2] and in reducing circulatory efficiency [3], [4].

Endothelial function can be evaluated non-invasively in brachial artery, with a technique known as flow mediated dilation (FMD) [5]-[8]. Basically, ultrasound images of the brachial artery are analyzed, before (baseline) and during an induced hyperemia by occlusion and release of a pneumatic cuff placed on the forearm. Increased post ischemia flow induces the release of nitric oxide (NO) in that artery, resulting in local vasodilation. The FMD is estimated as the percentage of change between maximum diameter achieved during hyperemia with respect to baseline diameter. Despite its widespread use, it is an expensive method and critically dependent on operator skills, limiting its application in everyday clinical practice [8], [9].

Pulse wave velocity (PWV), which is inversely related to distensibility, has attracted much interest in recent years as a measure of conduit artery stiffness. Distensibility, however, is influenced by dynamic changes in vascular tone as well as by chronic structural changes in the artery wall and may thus be used to measure acute changes in vascular tone. PWV has indeed been shown to be acutely influenced by constitutively released NO. Therefore, the PWV may be used to analyze the acute changes in vascular smooth muscle tone.

In this work we propose to analyze the use of carotid-radial PWV, as a tool to assess endothelial dynamics using the basic concept of the flow mediated dilation. This approach would evaluate endothelial dysfunction, in an independent way from the operator using low cost and widely available devices. The aims are: 1) characterize the beat to beat carotid-radial PWV temporal profile in response to transient ischemia in the forearm and 2) analyze the repeatability of the proposed technique.

2. Material and methods

2.1. Subjects

Ten healthy students (6 female, 22±2 years, 64±19 Kg, 1.7±0.1 m) from the Medicine School (Universidad de la República, Uruguay) were invited and agreed to participate in the non-invasive study. Measurements were performed in the morning, in a quite room, with temperature controlled at 21-23 ºC. Caffeine, alcohol and vitamin C ingestion, as well as strenuous exercise were avoided prior to the examination. The study protocol was approved by the ethics committee and all the participants gave informed consent.

2.2. Study protocol and recordings

All the procedures agreed with international consensus to evaluate the endothelial function through FMD [12]. The subjects remained rested in supine position for 15 minutes before starting the study. Right brachial blood pressure was measured with a sphygmomanometer every two minutes throughout the study, recording the systolic pressure (SP) and the diastolic pressure (DP). A pneumatic cuff was placed in the left forearm to induce a transient ischemia of 5 minutes (see figure 1).

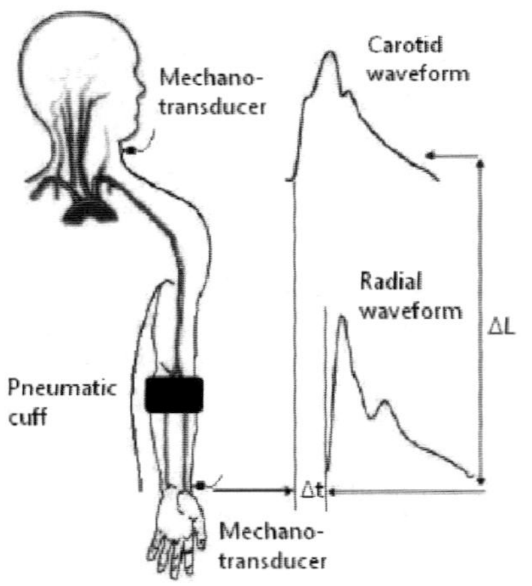

Figure 1. Instrumental employed to realize transient ischemia (pneumatic cuff) and the measurement of carotid-radial pulse wave velocity (mechano-transducers).

The carotid-radial PWV was measured using mechano-transducers placed simultaneously on the skin over the left carotid and radial arteries (see figure 1), using a similar device previously described [13]. Upon completion of each study, signals were analyzed using a software previously developed in our laboratory that allows beat to beat PWV calculation [13]. The PWV was obtained from the ratio of the distance between the sensors (ΔL) and the temporal delay (Δt) between carotid and radial waveforms. The obtained PWV recordings were: before cuff inflation (baseline) and during 10 minutes after cuff release (post-occlusion status).

In order to analyze the repeatability of the technique, in a subgroup of 5 randomly selected subjects the study was repeated one hour later under the same conditions. All the measurements were done by the same trained operator.

2.3. Data analysis

To analyze the endothelial dynamics, the relative carotid-radial PWV change was quantified in reactive hyperemia (PWV_{RH}), defined in equation (1):

$$PWV_{RH} = \frac{PWV_{Basal} - PWV_{Min}}{PWV_{Basal}} \cdot 100 \quad (1)$$

where PWV_{Basal} represents the average PWV of at least 10 beats of baseline recording prior to the occlusion and PWV_{Min} is the minimum value of PWV reached during hyperemia. To analyze the temporal evolution of carotid-radial PWV, after transient ischemia, the response of the artery of each subject was normalized, with respect to baseline.

For the repeatability study the coefficient of repeatability (CR) was calculated from the absolute difference (D_i) between the n=5 values of PWV_{HR} obtained in studies 1 and 2, as defined in equation (2):

$$CR = 2 \cdot \sqrt{\frac{\sum D_i^2}{n}} \quad (2)$$

We also evaluate the coefficient of variation (CV) as the ratio between the standard deviation of the two measures and the average.

Data are expressed as mean ± standard deviation. Comparison between variables were analyzed by paired Student t test or ANOVA as appropriate. A value of $p<0.05$ was considered statistically significant.

3. Results

Baseline SP and DP values were 118±11 y 74±9 mmHg respectively. There were no significant changes observed in SP and DP, between the records at baseline, occlusion and post-occlusion. Figure 2 shows the carotid-radial PWV temporal profile normalized with respect to baseline of each student, for the entire group. The initial time of the profile represent the moment when the cuff is deflated and the first radial beat is displayed. The minimum time (t_{min}), calculated as the time at which the PWV is equal to PWV_{Min} after reactive hyperemia, was 59±31 seconds. The recovery time (t_{rec}), calculated as the time at which the PWV reaches a maximum, for at least 5% of baseline, was 533±65 seconds. PWV_{Min} value was significatively lower than PWV_{Basal} (664±137 vs. 830±107 cm/s; $p<0.05$). The percentage change in PWV during the hyperemia compared to baseline was PWV_{RH}=20.7±9.2 %. The individual values obtained before and after transient ischemia, reflected in PWV measurements, are showed in table 1.

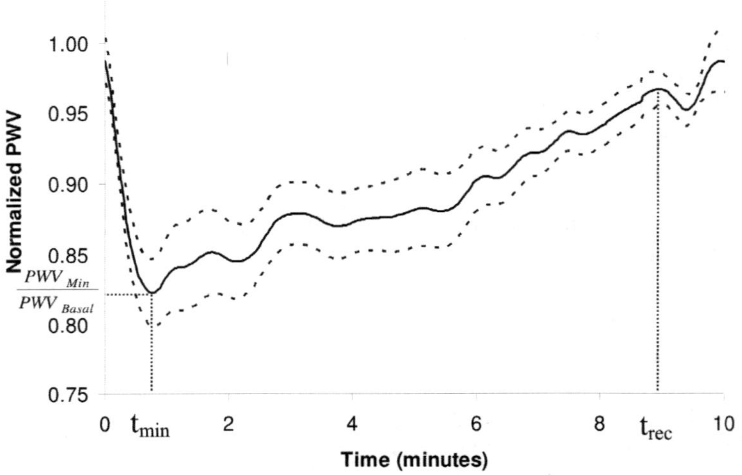

Figure 2. Average carotid-radial PWV temporal profile, normalized with respect to baseline value, calculated during 10 minutes after transient ischemia, for the 10 subjects. Solid line: average. Dotted line: standard error of the mean.

There were no significant changes in SP and DP, or heart frequency between the considered studies for the repeatability test, ensuring that endothelial dynamics was evaluated in stable hemodynamic conditions. There were no significant differences between PWV_{RH} in study 1 ($PWV1_{RH}$=15.1±2.7 %) and study 2 ($PWV2_{RH}$=14.1±2.3 %).

Table 1. Derived parameters from carotid-radial PWV before and after transient ischemia.

Subject	Pwv_{Basal} (cm/sec)	Pwv_{Min} (cm/sec)	Pwv_{HR} (%)	t_{min} (sec)	t_{rec} (sec)
1	860±41	715	16.9	47	609
2	850±37	734	13.6	42	612
3	784±34	679	13.4	46	508
4	620±10	399	35.6	98	540
5	897±18	613	31.7	37	512
6	979±45	854	12.8	129	468
7	949±25	769	19.0	55	526
8	795±20	654	17.7	27	404
9	716±33	476	33.5	57	597
10	858±29	748	12.8	51	552

We obtained a CR of 4.8% and a CV of 9.7%. Individual values for the subgroup of 5 subjects considered for the repeatability study are shown in table 2.

Table 2. Results of the repeatability study.

Subject	$Pwv1_{HR}$ (%)	$Pwv2_{HR}$ (%)
1	16.9	16.2
2	13.6	12.8
3	13.4	16.1
6	12.8	10.9
7	19.0	14.9

Figure 3 shows the carotid-radial temporal profiles, normalized with respect to baseline, for the first and the second measurement performed one hour later. It is noted that the two temporal evolutions of PWV show the same trend, i.e. both reach a minimum value around the minute, with a very similar relative decrease, and then begin to increase to recover its initial value.

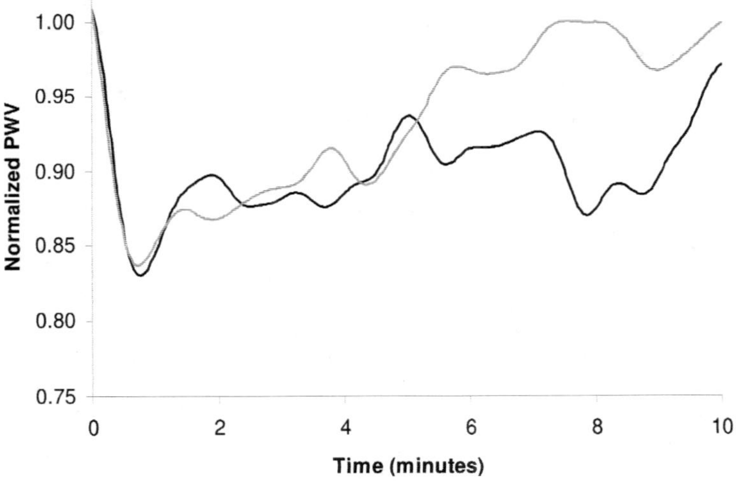

Figure 3. Result of the repeatability study. Response of carotid-radial PWV normalized with respect to baseline value calculated during 10 minutes after transient ischemia, for one subject. Black: first measurement. Gray: second measurement, one hour later.

4. Discussion

This paper presents a new methodology in order to evaluate no invasively endothelial dynamics. This technique is based on the beat to beat carotid-radial PWV measurement, before and during reactive hyperemia induced by the occlusion and release of a cuff placed on the forearm.

From the analysis of carotid-radial PWV temporal profiles, there is a clear decrease of PWV during hyperemia compared to baseline and subsequent recovery after several minutes. From the Moens Korteweg relation, it is possible to relate the square of pulse wave velocity, directly to the product of the elastic modulus and wall thickness, and inversely with the product of the arterial diameter and the density of blood [13]. Therefore, the sharp decline in PWV observed during hyperemia may be related mainly to an increase of diameter and/or a decrease of elastic modulus. In this sense, it was shown that local activation of vascular smooth muscle by nitroglycerin, causes isobaric increases of diameter and isobaric decreases of elastic modulus [14]. We found that the percentage of change between PWV_{Min}

achieved during hyperemia compared to baseline was near 21%. These findings are consistent with an other work that applying a similar methodology, compares the technique of FMD with the humerus-radial PWV measurement [15]. In that work, the relative reduction of PWV respect to baseline values evaluated in 17 healthy subjects of 33±11 years, was 14±5 %. Due to the fact that in healthy individuals, diameter changes assess by the ultrasound FMD technique yields values on the order of 7-9 % [7], [10], one could infer that the present methodology is more sensitive. Similarly, an interesting point to emphasize is that although the time to reach the minimum value for PWV is similar to the time to reach the maximum dilation in FMD technique (both within the minute of cuff release), recovery times are significantly different. The recovery time here found was about 9 minutes, while in the FMD technique was around 3 minutes [12], [16]. In a recent study involving FMD, using an automatic method that allows the continuous diameter detection, a better sensitivity was achieved analyzing the ascendant speed of the profile instead of extreme values [17]. Perhaps our methodology will require in a future, to incorporate a descendent profile rate evaluation, in order to analyze in more detail the recovery phase. Therefore, in future studies of FMD in addition to the temporal signal of diameter, it would be very interesting to obtain simultaneously the evolution of the elastic modulus. An alternative could be the simultaneous measurement of diameter and PWV, using the here proposed methodology.

Although PWV measurement is a simple and widely used technique to assess arterial stiffness, the approximate determination of the separation distance between sensors required for its calculation, result in an under or overestimation of the true value. In this work analyzing the relative changes in PWV, each individual was self-control. Moreover, PWV is strongly related to blood pressure. An increase in blood pressure level, will increase PWV. Therefore one might think that the decrease of PWV during hyperemia was actually due to a passive phenomenon, as a response of a decrease in blood pressure. If this were the case, FMD studies should not show a vasodilator response. On the contrary, if during hyperemia pressure level increases, a decrease of PWV is not expected. Therefore, the changes found during hyperemia can not be entirely attributed to a pressure phenomenon.

It is known that endothelial function studies based on analysis of ultrasound images are extremely variable, with coefficients of variations up to 40% [9]. The use of automated methods have noticeably reduce the variability, presenting CV values of 16% for one hour repeated studies, and 18% for one month repeated studies [16]. In the present study using the beat to beat PWV technique, the resulting repeatability coefficient for PWV_{RH} was less than 5% and the CV less than 10%. It is to note, that future studies involving inter-operator variability are required.

Finally, the results were obtained in a small sample for a young and healthy population. It is therefore necessary to perform other studies, including a higher number of patients of different ages and with different pathologies that produce endothelial dysfunction, to confirm the results here found. These issues, as well as the possibility to compare the results with simultaneous recordings of other techniques such as FMD, are currently being planned by our working group for the near future.

5. Conclusion
Through analysis of beat to beat carotid-radial PWV it was possible to characterize in healthy subjects the temporal profiles and to analyze the acute changes in response to a transient ischemia. The results show that the technique has a high sensitivity and repeatability.

Acknowledgments
This research was supported in part by grant from ECOS-Sud (Argentine/France) project N° A06S02 and by Agencia Nacional de Investigación e Innovación (FCE-2007-635, FCE-2007-638 and PRSCT-008-020) from Uruguay.

Journal of Physics: Conference Series **313** (2011) 012009 doi:10.1088/1742-6596/313/1/012009

References

[1] Ross R 1993 The pathogenesis of atherosclerosis: a perspective for the 1990s *Nature* **362** 801-9

[2] Alexander R W and Dzau V J 2000 Vascular biology: the past 50 years 2000 *Circulation* **102** 112-6

[3] Griffith T M, Edwards D H, Davies R L, Harrison D J and Evans K T 2000 NO coordinates the behavior of vascular resistance vessels *Nature* **329** 442-5

[4] Henderson A H 1997 Endothelial dysfunction: a reversible clinical measure of atherogenic susceptibility and cardiovascular inefficiency *Int. J. Cardiol.* **62** 543-8

[5] Laurent S, Lacolley P, Brunel P, Laloux B, Pannier B and Safar M 1990 Flow-dependent vasodilation of brachial artery in essential hypertension *Am. J. Physiol.* **258** 1004-11

[6] Anderson E A and Mark A L 1989 Flow-mediated and reflex changes in large peripheral artery tone in humans *Circulation* **79** 93-100

[7] Celermajer D S, Sorensen K E, Gooch V M, Spiegelhalter D J, Miller O I, Sullivan I D, Lloyd J K and Deanfield J E 1992 Non-invasive detection of endothelial dysfunction in children and adults at risk of atherosclerosis *Lancet* **340** 1111-5

[8] Sorensen K E, Celermajer D S, Spiegelhalter D J, Georgakopoulos D, Robinson J, Thomas O and Deanfield J E 1995 Non-invasive measurement of human endothelium dependent arterial responses: accuracy and reproducibility *Brit. Heart J.* **74** 247-53

[9] Malik J, Wichterle D, Haas T, Melenovsky V, Simek J and Stulc T 2004 Repeatability of noninvasive surrogates of endothelial function *Am. J. Cardiol.* **94(5)** 693-6

[10] Ramsey M W, Goodfellow J, Jones C J H, Luddingtoin L A, Lewis M J and Henderson A H 1995 1995 Endothelial control of arterial distensibility is impaired in chronic heart failure *Circulation* **92** 3212-19

[11] Kinlay S, Creager M A, Kukumoto M, Hikita H, Frang J C, Selwyn A P and Ganz P 2001 Endothelium-derived nitric oxide regulates arterial elasticity in human arteries in vivo *Hypertension* **38** 1049-53

[12] Correti M C *et al* 2002 Guidelines for ultrasound assessment of endothelial-dependent flow-mediated vasodilation of the brachial artery: a report of the International Brachial Artery Reactivity Task Force *J. Am. Coll. Cardiol.* **39** 257-65

[13] Graf S 2008 Detección no invasiva de las alteraciones arteriales inducidas por la ateroesclerosis y la hipertensión arterial. PhD thesis. Facultad de Medicina. Universidad de Buenos Aires. Exp. No 506.644/01, Res: No 1727

[14] Bank A J, Kaiser D R, Rajala S, Cheng A 1999 In vivo human brachial artery elastic mechanics: effects of smooth muscle relaxation *Circulation* **100** 41-7

[15] Naka K K, Tweddel A C, Doshi S N, Goodfellow J and Henderson A H 2006 Flow-mediated changes in pulse wave velocity: a new clinical measure of endothelial function *Eur. Heart J.* **27** 302-9

[16] Craiem D, Chironi G, Gariepy J, Miranda-Lacet J, Levenson J and Simon A 2007 New monitoring software for larger clinical application of brachial artery flow-mediated vasodilatation measurements *J. Hypertens.* **25(1)** 133-40

[17] Chironi G, Craiem D, Miranda-Lacet J, Levenson J, Simon A 2008 Impact of shear stimulus, risk factor burden and early atherosclerosis on the time-course of brachial artery flow-mediated vasodilation *J. Hypertens.* **26(3)** 508-15

Journal of Physics: Conference Series **313** (2011) 012010 doi:10.1088/1742-6596/313/1/012010

CardioGRID: a framework for the analysis of cardiological signals in GRID computing

Juan Francisco Garcia Eijo[1], Marcelo Risk[1], Francisco Prieto Castrillo[2], Cesar Suarez Ortega[2], Maria Boton Fernandez[2], Alfonso Pardo Diaz[2], Manuel Rubio del Solar[2] and Raul Ramos Pollan[2]

[1]Laboratory of Complex Systems, Department of Computer Science, FCEyN. University of Buenos Aires, Argentina.
[2]Extremadura Research Centre for Advanced Technologies. Research Centre for Energy, Environment and Technology (CETA-CIEMAT). Trujillo, Spain.

E-mail: jfgarcia@dc.uba.ar - mrisk@dc.uba.ar

Abstract. The present paper describes the development of the CardioGRID framework into the GRID infrastructure. The core GRID services; Workload Management System (WMS), Data Management System and Grid Authentication have been implemented. Additionally, a web-based tool -the CardioGRID portal- has been developed to facilitate the user interaction with the GRID. As a result, the user is able to process the electrocardiogram (ECG) signals obtained form a portable data acquisition device and to process it on the GRID. Once the CardioGRID portal is prompted and the user identity is verified through a digital X.509 certificate, the operator may either upload new raw ECG data to the GRID Storage Elements or use already stored data. Then, subsequent analytics from these data are performed as GRID jobs and relevant medical quantities are derived through middle-ware job retrieval mechanism. In summary in this paper was described the development of a medical GRID based system, and its integration to an existing platform for Digital Repositories Infrastructure.

1. Introduction

1.1. e-Health Technologies and GRID

The group of technologies used in e-Health are mainly electronic and communications processing of medical data, in order to assist health care, such as telemedicine, health care information systems, electronic medical records and mobile devices for health care, among others [1].

Telemedicine is the use of devices and communications systems, where physiological measurements in patients are transferred using Internet, telephone, or two way digital TV systems, between the patients location to a health care facility; applications of telemedicine includes access to patients in remote locations, or simply access to patients in their home for routine control [2].

Health care information systems applies computers resources, such as devices, storage, retrieval, and programs, to implement clinical guidelines and processes. These guidelines and processes involve the architecture of medical records to manage patients information, medical decision support, standards of medical data, images and signals (HL7 and DICOM) [3].

Using mobile devices for health care -also known as mHealth- comprises the interaction of different mobile devices and technologies, such as mobile and smart phones, PDAs, combinations of all them, together with physiological sensors to measure patients signals [4, 5].

mHealth can be though as telemedicine as far as it is the technology which provides patients data, both from sensors and demographic data. The mHealth technology seems to be an alternative for low income countries [6] where mobile phones and services are accessible to most people by providing the last mile connection for large telemedicine networks [5].

On the other hand, distributed computing technologies (and more specifically, GRID computing) are becoming more and more popular among researchers. In this regard, the efforts of European Commission Projects, such as Enabling Grids for e-Science (EGEE) [7] or E-Science Grid Facility for Europe and Latin America (EELA-2) [8] along with the National Grid Initiatives [9] are bridging the gap between the end user and GRID.

Three main features characterize the GRID infrastructure; computing and storage power, security and collaboration. Whereas in a High Performance Computing environment (HPC), the computing power plays an essential role, it is in the possibility of collaboration where High Throughput Computing (i.e. GRID computing) finds its major strength.

GRID computing provides a powerful alternative to handle large amounts of data and processing, therefore, it has potential to be used in medical research, where is common clinical trials involving large patients databases, and physiological and genomes modeling, which needs intensive computer processing [10, 11].

1.2. Cardiological Signals

The cardiological signals such as the electrocardiogram (ECG), and the blood pressure (BP), show the electrical activity and the main mechanical effect of the heart, respectively; from the ECG signal can be derived the heart rate (HR), detected measuring the interval between adjacent R waves from the ECG [12].

The HR variability (HRV) is an indicator of the beat-to-beat modulation by the autonomic nervous system (ANS) [12, 13]. The diseases which affects the ANS, such as diabetes, leads to a an impaired regulation of the heart rate, this can be detected analyzing the HR variability [13].

The length of ECG recordings varies from minutes to several hours, in the case of Holter tests the ECG is recorded during 24 hours [13]. In the Figure 1 can be observed the HR obtained from a healthy patient, during 24 hours, showing large changes the day of the test; assuming a mean heart rate of 70 beats per minute, the total number of beats can be more than 100,000 beats for 24 hours.

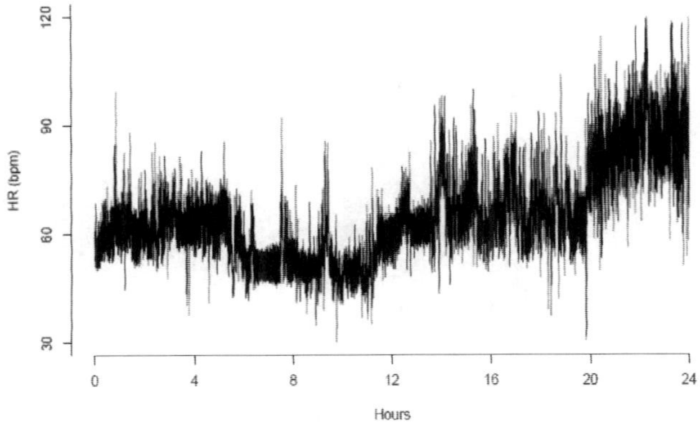

Figure 1. HR versus time for a Holter test of 24 hours.

Journal of Physics: Conference Series **313** (2011) 012010 doi:10.1088/1742-6596/313/1/012010

1.3. Objectives
The CardioGRID framework objectives were the development of a mobile device for ECG and data recording, and the development of a GRID computing system for the processing of ECG, the calculation of HR, the classifications of heart beats, and both time and frequency domain of the HRV.

2. GRID Implementation for the CardioGRID Portal
2.1. The CardioGRID Framework
The CardioGrid framework comprises three basic elements, as shown in Figure 2.

- Mobile device for ECG recording and data patient entry (smart phone with ECG module),
- Web portal to transfer ECG and data from patients,
- Grid infrastructure for ECG processing.

The Portal possesses two possible use cases:

(i) The user, either a physician or a technician, who is in a rural zone where the patients do not have access to a full-equipped health center, through a smart phone and the ECG acquisition module obtains the ECG patients. When the user is returns to a urban zone with Internet access, will be able to connect to the Portal and transfer the ECGs to process them.

(ii) The patient ECG is stored in the PC of the user. In this case, the user, access via Internet to the portal, uploads the patient personal data and attaches the ECG that desires to process on the GRID infrastructure. Once that it was delivered the user could consult the process state, and obtain the results when it is finished.

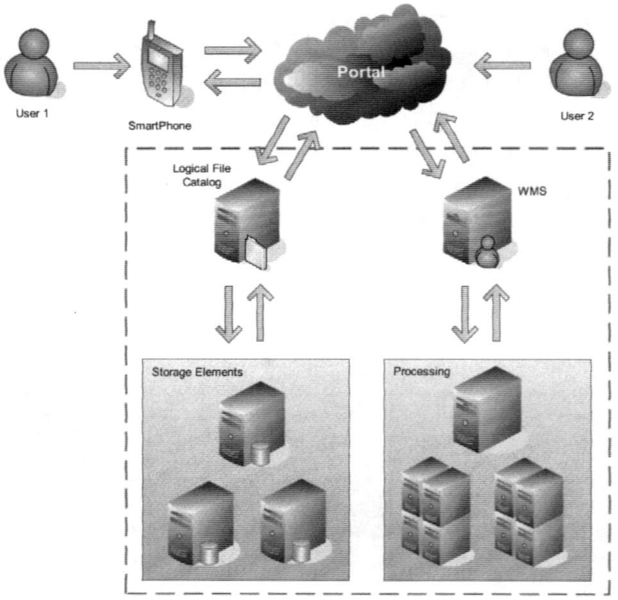

Figure 2. CardioGrid framework scheme, showing the three basic elements: mobile device using a smart phone, web portal and Grid infrastructure.

The Figure 3 shows an smart phone acquiring ECG from a patient, the connection between the ECG module and the smart phone is using bluetooth connectivity.

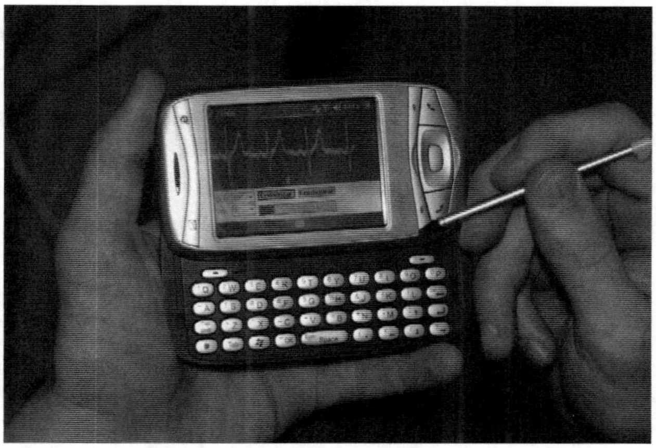

Figure 3. Smart phone acquiring ECG from a patient.

2.2. The ECG processing on the GRID.
Once that the user completes the patient data and attaches his ECG satisfactorily from the Portal, the ECG with file extension (.dat) is transfered via Internet to a storage element (SE) of GRID infrastructure and its physical direction is stored in a Logical File Catalog (LFC) as it is shown in Figure 4.

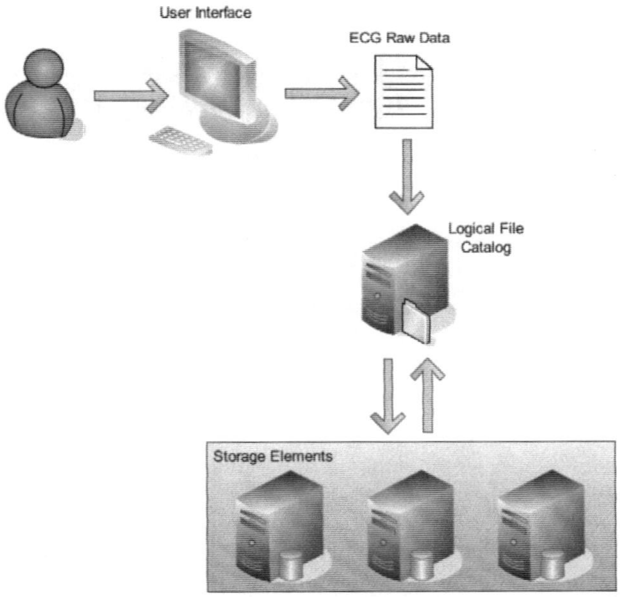

Figure 4. ECG raw data file (.dat), which is transfered to the SE of the Grid infrastructure.

Then the Workload Management System (WMS) selects an available Computing Element (CE) that adjusts to our work and sends to process the appropriate ECG. It is process using a special algorithm developed in C++ for wave detection R [13]; the transformation results are temporary stored a Worker Node (WN), as it is shown in Figure 5.

The ECG detection file (.txt) is processes with the ECG data file (.dat) in order to improve the detection rate.

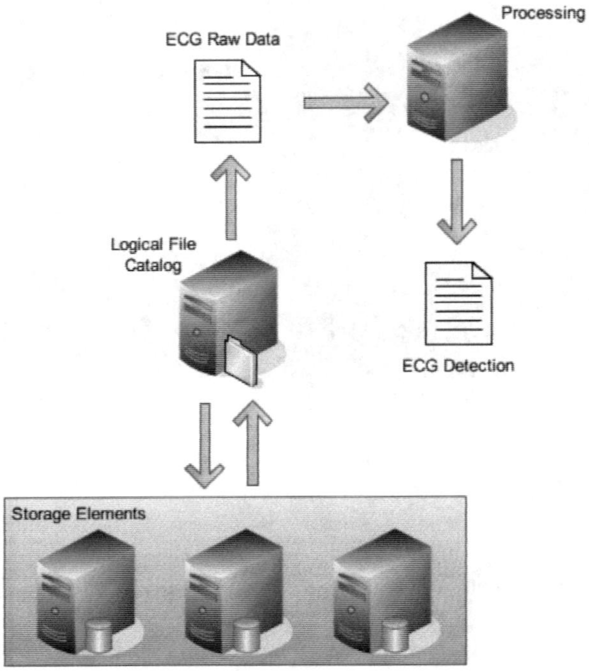

Figure 5. ECG raw data file processing. The results (,txt) is stored temporally in the WN.

The result of this step is the RR (intervals between adjacent R waves of the ECG), as we can see in Figure 6. This file represents a HR plot versus time, this could be seen in Figure 1.

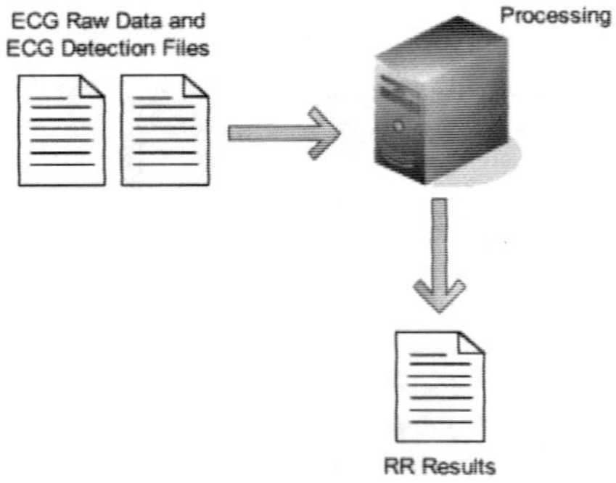

Figure 6. Combined processing of ECG Detection files and ECG Raw Data files, in order to refine the beat detection.

The last step consists in the statistic process of the interval RR file (.csv) to obtain the results file.

Once that the process on the GRID is complete, the user could choose whether to download the results to his PC or to just visualize them in the web (Figure 7).

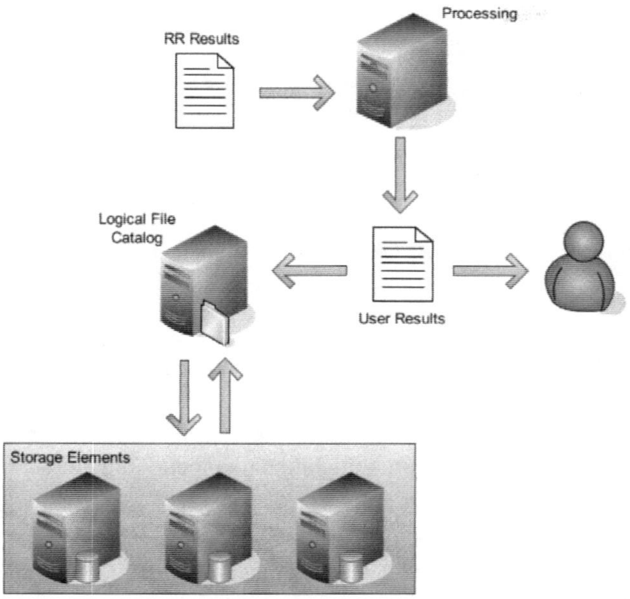

Figure 7. Statistical processing of the RR intervals file (.csv).

3. Conclusion

The present paper describes the development of a GRID based system, suited for an specific medical application; this kind of applications requires secured access, confidentiality of data, data replication, and fault tolerance. All of the above specifications can be filled using a GRID computing based infrastructure.

In the case of secured access, users can benefit using GRID certificates, based on the X.509 specification.

The confidentiality of the patients data is addressed by the use of encrypted transfers, in this project we were the first using the secure copy service implemented in gLite.

This mechanism allows several groups to work on the same project, and more importantly to share resources across the Grid securely. Data replication and fault tolerance are easy to implement in a GRID infrastructure, since by definition these systems are distributed, allowing multiple copies of the data at different physical machines, and therefore increase the fault tolerance.

The benefits for users accessing via web portal, is a very easy access to the power of the Grid, since they dont need any Grid computing knowledge.

The algorithms implemented in this paper dont need a high power of processing, however the advantage using Grid computing is the distribution of several simultaneous processes along the Grid, instead a single computer to process all of them in a queue.

More advanced algorithms, such as neural networks among others, could be easily added to a Grid based infrastructure, and rapidly available to all users.

4. Future Work

The work presented in this paper settled the basis for the development of a GRID-based repository framework for ECG analysis.

Basically, a Digital Repositories Infrastructure (DRI) constitutes a platform to host and manipulate arbitrary data structures on the GRID [14]. In this regard, the CardioGRID framework can benefit from the advantages of a full GRID-based tool which exploits the sharing and computational capabilities of the GRID. Hence, new communities could deploy ECG repositories on the DRI engine, perform sophisticated analyses on them and share results amongst researchers.

Here, the main architecture layers are shown:

- A user interface (UI) allowing users to navigate into federated data space.
- Grid storage elements to hold the main bulk of the digitalized data.
- A federated database for storing meta information about the data stored in the grid.

In this regard, the ECG work shown in this article could be adapted to a DRI repository by defining a ECG data model. Then, in an automatic way, the ECG data and its processing mechanisms benefit directly from the Grid Features through the DRI framework.

Acknowledgments

The authors are grateful to Diego Fernndez Slezak, Pablo Turjanski and Guillermo Marshall for the collaboration in the current work; to the CETA-CIEMAT and the EELA-2 project for their support in the gridification of this project, and to Microsoft Research for their support in the development of the mobile device.

MR and JFGE want to acknowledge the Department of Computer Science, Faculty of Exact and Natural Sciences, University of Buenos Aires, and the National Research Council of Argentina (CONICET), Argentina.

References

[1] J. Fingberg, M. Hansen, M. Hansen, H. Krasemann, L. L. Iacono, T. Probst, and J. Wright, "Integrating data custodians in ehealth grids - a digest of security and privacy aspects." in *GI Jahrestagung (1)*, 2006, pp. 695–701.

[2] A. C. Norris, "Essentials of telemedicine and telecare," *John Wiley and Son*, p. 177, 2002.

[3] "Digital imaging and communications in medicine (dicom) part 1: Introduction and overview," National Electrical Manufacturers Association, 300 N. 17th Street, Rosslyn, Virginia 22209 USA, Tech. Rep., 2006.

[4] R. Istepanian, "Introduction to the special section on m-health: Beyond seamless mobility and global wireless health-care connectivity." in *IEEE Transactions on Information Technology in Biomedicine, 8(4)*, 2004, pp. 405–413.

[5] "mhealth for development: The opportunity of mobile technology for healthcare in the developing world." United Nations Foundation, Vodafone Foundation., Tech. Rep., 2009.

[6] W. A. Kaplan, "Can the ubiquitous power of mobile phones be used to improve health outcomes in developing countries?" *Globalization and Health*, vol. 2, pp. 9+, May 2006. [Online]. Available: http://dx.doi.org/10.1186/1744-8603-2-9

[7] E. Laure and B. Jones, "Enabling grids for e-science: The egee project," Tech. Rep. EGEE-PUB-2009-001. 1, Sep 2008.

[8] L. N. Ciuffo, "The eela project," May 2007, presented at IST-Africa Conference 2007, Maputo - Mozambique.

[9] "Spanish national grid initiative, retrieved on may 2009." [Online]. Available: http://www.e-ciencia.es

[10] S. Cheung, X. Chu, S. Xu, R. Buyya, and J. Tu, "A grid based e-research platform for clinical management in the human respiratory and vascular system," in *ACSW '07: Proceedings of the fifth Australasian symposium on ACSW frontiers*. Darlinghurst, Australia: Australian Computer Society, Inc., 2007, pp. 11–17.

[11] H. Gjermundrod, M. D. Dikaiakos, D. Zeinalipour-Yazti, G. Panayi, and T. Kyprianou, "Icgrid: Enabling intensive care medical research on the egee grid." Health Grid, 2007, p. 10.

[12] S. Akselrod, D. Gordon, J. B. Madwed, N. C. Snidman, D. C. Shannon, R. J. C. J. B. Madwed, N. C. Snidman, D. C. Shannon, and R. J. Cohen, "Hemodynamic regulation: Investigation by spectral analysis," *American Journal of Physiology*, vol. 249, no. 18, pp. H867–H875, 1985.

[13] M. Risk, V. Bril, C. Broadbridge, and A. Cohen, "Heart rate variability measurement in diabetic neuropathy: review of methods." *Diabetes Technol Ther*, vol. 3, no. 1, pp. 63–76, 2001. [Online]. Available: http://view.ncbi.nlm.nih.gov/pubmed/11469709

[14] A. Calanducci, F. P. Castrillo, R. R. Pollan, and M. R. del Solar, "Enabling digital repositories on the grid," *Advanced Engineering Computing and Applications in Sciences, International Conference on*, vol. 0, pp. 45–50, 2008.

Feature Extraction on Brain Computer Interfaces using Discrete Dyadic Wavelet Transform: Preliminary Results

I Gareis[1], G Gentiletti[1], R Acevedo[1,2], L Rufiner[1,2]
[1] Laboratorio de Ingeniería en Rehabilitación e Investigaciones Neuromusculares y Sensoriales - Facultad de Ingeniería, Universidad Nacional de Entre Ríos, Argentina.
[2] Centro de I+D en Señales, Sistemas e Inteligencia Computacional - Facultad de Ingeniería y Ciencias Hídricas, Universidad Nacional de Litoral, Argentina.

racevedo@bioingenieria.edu.ar

Abstract. The purpose of this work is to evaluate different feature extraction alternatives to detect the event related evoked potential signal on brain computer interfaces, trying to minimize the time employed and the classification error, in terms of sensibility and specificity of the method, looking for alternatives to coherent averaging. In this context the results obtained performing the feature extraction using discrete dyadic wavelet transform using different mother wavelets are presented. For the classification a single layer perceptron was used. The results obtained with and without the wavelet decomposition were compared; showing an improvement on the classification rate, the specificity and the sensibility for the feature vectors obtained using some mother wavelets.

1. Introduction

The technique most commonly employed to improve the Signal to Noise Ratio (SNR), and estimate the Evoked Potential signal (EP), is coherent averaging. It consists on applying successive stimuli and average the electrical activity registered synchronized with the time instant in which the stimulus is applied.

Nevertheless coherent averaging is of massive use, it presents some issues that make desirable the finding of alternatives to this technique. One of these issues is the time required to achieve an accurate estimation of the EP, another one is that it assumes that the EP remains stationary from trial to trial and that the noise is white with zero mean, which is not completely true in general. As a consequence the result obtained from the coherent averaging is a bad estimation of the EP signal [1].

On the other hand there are situations in which it is not necessary to estimate the signal but rather to detect if it is present or not. This is the case of Brain Computer Interfaces (BCI). There are many BCI paradigms that can be initially classified in two groups: a) Invasive: in this case the signals are provided by intracranial or intracortical registers and b) Non invasive: using surface registers, from which the most common are the ones based on electroencephalography (EEG). To this last group belongs the paradigm based on Event Related Potentials (ERP) [2].

ERP are potentials of latencies larger than 100 ms, and their manifestation depends of psychological and cognitive processes. The most studied ERP is the P300 signal. When visual, auditory or somatosensory infrequent or particularly meaningful stimuli are mixed with frequent stimuli, the first evoke a potential over the parietal cortex with a positive peak close to the 300 ms that

Journal of Physics: Conference Series **313** (2011) 012011 doi:10.1088/1742-6596/313/1/012011

can be registered on the EEG. To estimate the ERP signal it is necessary to improve the initial SNR by means coherent averaging.

Donchin et al [3] were the first to use the p300 on a BCI with visual stimuli organized by rows and columns over a character matrix that allowed its spelling. The most outstanding advantage of this technique is that it does not require the user to train, and on the same time it results natural on target selection tasks e.g. word spelling, direction selection, menu selection, among others. Many works suggest that P300 based BCIs can achieve a data communication rate of 25 bits/min working online [4].

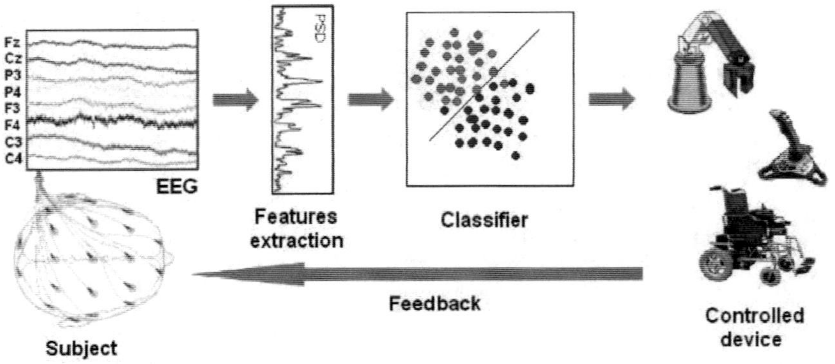

Figure 1. General architecture of a BCI for device control.

In Figure 1 is shown the general architecture of a BCI proposed by Millán et al [5], were its functional blocks are described, in which a subject controls a device, e.g. a wheelchair, a robotic arm or a joystick. The user can also check the device state trough feedback that allows him to acknowledge the result of his efforts to control it.

Recently, with the advance in algorithms on machine learning and technologies of digital processing, a part of the investigation on BCI is the exploration of feature extraction techniques and signal classification.

On a BCI like the one described the following blocks can be clearly distinguished:

a) Instrumentation block: is the acquisition and conditioning system for the electroencephalogram.

b) Feature extraction block: its objective is to generate a suitable representation of the original signal (pattern in the context of Pattern Recognition) that allows improving the performance of the classifier.

c) Classifier block: its function is to identify to which class does a pattern belongs to.

In this particular problem the last two blocks are a pattern recognition system where the classification problem to solve has two possible classes: registers with ERP and registers without ERP. The objective of this work is to evaluate different feature extraction alternatives to detect the ERP signal in BCI minimizing the time employed, and maximizing as well the accuracy, sensitivity and specificity of the method. We present here some preliminary results of applying a signal decomposition technique based on dictionaries of time-frequency atoms, which are generated using Wavelet multiresolution decomposition.

2. Methodology

2.1. ERP registers

In this work there were used registers obtained with amplifiers Grass® model 8-18-36, the acquisition parameters of the EEG signals are displayed on table I. The BCI2000 software developed in the Wadsworth Center (Albany, New York) [6] was used.

Table 1. EEG acquisition parameters.

Parameter	Description
Channels	Fz, Cz, Pz, Oz, C3, C4, M1 (reference) y M2 (ground)
Sample Frequency	1024 Hz

To minimize the amount of information presented to the classifier successive decimations were performed over the temporal patterns, with the objective of setting the amount by which the number of samples could be reduced without reducing the performance of the classifier. Care was taken to avoid the aliasing by applying a Chebyshev filter. The decimations were made by multipliers of two for convenience.

For the stimulation, the Donchin speller [3] was used, although the alphabet characters were replaced by icons that control the movement of a wheelchair [7].

2.2. Feature Extraction

There are many alternatives for the feature extraction block, some more suitable than others depending on the signal and the features that are intended to be emphasized. In this work we propose to use approximation methods of approximation based on dictionaries of discrete signals like the used in [8],[9]. In these methods the signal of interest is considered as an element of a signal space that can be represented in terms of a dictionary or base that remark meaningful features. In this context the signal $\mathbf{x} \in \mathrm{R}^{N}$ can be expressed as a function of an appropriate orthogonal base using the equation $\mathbf{a} = \boldsymbol{\varphi}.\mathbf{x}$ where $\mathbf{a} \in \mathrm{R}^{N}$ is the vector expressed in the new base and $\boldsymbol{\varphi} \in \mathrm{R}^{NxN}$ is a matrix which columns are the elements of the new base, also called atoms [8],[9]. In this work the vector **a** was built by means of a multiresolution decomposition produced by the Discrete Dyadic Wavelet Transform (DDWT) using mother wavelets of different families which were selected by their "visual" resemblance with the P300 peak. In this process, the signal is separated in a portion of high frequencies (called detail) and a portion of low frequencies (called approximation), each with half of samples of the original signal, which represent the first level of decomposition. The decomposition in each posterior level is made only over the approximation of the previous one, obtaining another detail and approximation portions, with half the samples of the previous approximation.

Previous to the application of the DDWT the registers (signals) were filtered using a low pass Chebyshev digital filter of order eight with a cutoff frequency that depended on the subsequent subsampling, after this the registers were subsampled. After the DDWT different levels of details were selected with the approximation, with these coefficients there were formed vectors, one corresponding to each trial. These vectors formed the groups of patterns to present to the classifier. The coefficient groups were selected taking into account that the ERPs are formed by low frequency components, therefore the approximations were always kept and the amount of detail levels used were varied, always keeping the details of the bigger scales and adding detail coefficients of the smaller scales to be able to define which of the detail scales have useful information for the classification process.

2.3. Classifier

Assuming that the problem classes are linearly separable a single layer perceptron was used.

2.4. Performance

The indices chosen to evaluate the performance of classification were the accuracy, sensitivity and specificity, defined as follows:

$$\text{Accuracy} = \frac{\text{Correctly classified patterns}}{\text{Total of patterns}} \qquad \text{Sensibility} = \frac{\text{TP}}{\text{TP}+\text{FN}} \qquad \text{Especificity} = \frac{\text{TN}}{\text{TN}+\text{FP}}$$

Journal of Physics: Conference Series **313** (2011) 012011 doi:10.1088/1742-6596/313/1/012011

where TP are true positives values, TN are true negatives values, FP are false positives values and FN are false negatives values.

3. Results

3.1. Registers

In Figure 2 are shown the 300 trials averages ERP obtained in simultaneous registers on positions F_z, C_z, P_z, O_z, C_3 y C_4 in their respective positions. It can be seen that the morphology of ERP depends on the position of the electrodes on which the signal is acquired, and that in some positions it manifests better than in others. From the six were selected two channels on which the P300 peak is manifested with a different polarity, this were channels C_z and O_z. On them it was evaluated how does the amount of trials to average influence the time frequency analysis.

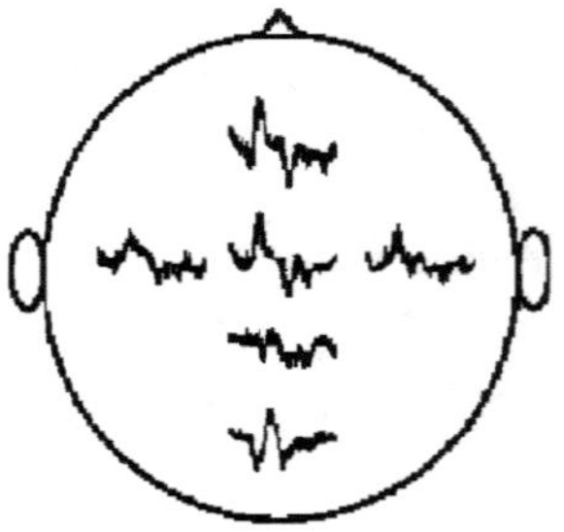

Figure 2. ERP waveform in different positions of the scalp.

Figure 3 shows the scalograms that result from applying Continuous Wavelet Transform (CWT), using Biorthogonal 3.9 as mother wavelet, in averages of 10, 5 and 1 trials to patterns with and without ERP from channels Cz and Oz. It can be seen that the scalograms show differences between the patterns with and without ERP, even though it is clearer on channel Oz. This situation is maintained noticeably for averages of up to three trials (not shown in this work) for channel Oz. Due to these we worked with data from channel Oz.

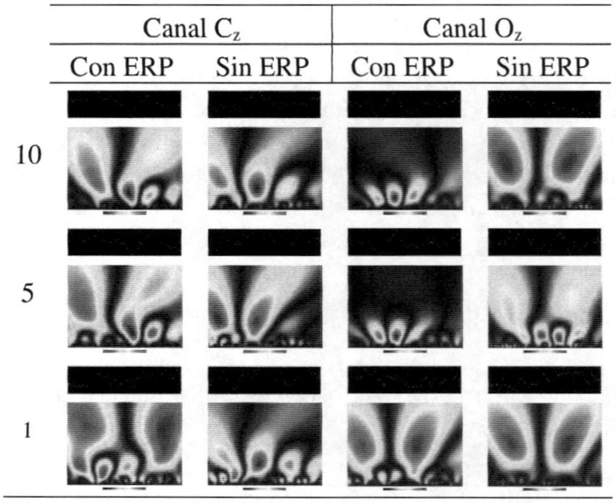

Figure 3. Scalograms of C_z and O_z channels with different amount of averaged trials

In Figure 4 the averages obtained from registers corresponding to channel Oz are shown; on blue line the average of tree hundred trials with ERP while on red line is displayed the average of tree hundred trials without ERP.

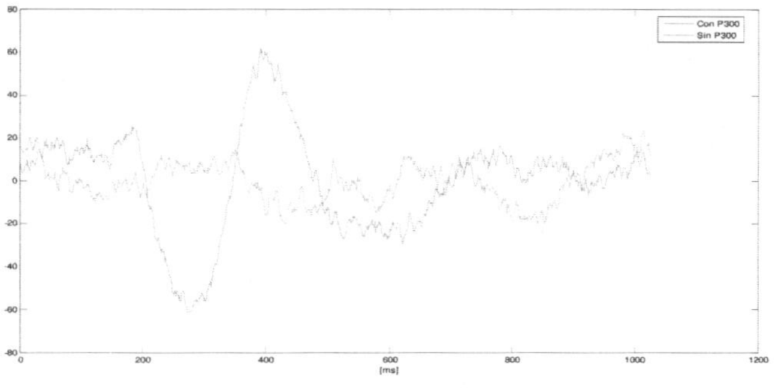

Figure 4. Averages with and without ERP on channel Oz.

3.2. Subsample

Table II summarizes the performance presenting single trial temporal decimated registers to the classifier. It can be seen that systems performance was maximum when decimating by four. Because of that decimation by four was set to be used previously to the application of the DDWT.

Table II. Systems performance for different decimation levels

Decimation	Accuracy	Sensitivity	Specificity
No decimation	63 %	0,7	0,56
x 2	63 %	0,72	0,54
x 4	64,75 %	0,755	0,54
x 8	61,75 %	0,69	0,54
x 16	62 %	0,7	0,53
x 32	62 %	0,69	0,55

3.3. ERP detection

Considering the results obtained in the time-scale analysis previously described, and the performance of the system for the decimated temporal patterns, sets of temporal patterns that consisting of single trials corresponding to the O_z channel were generated, all decimated four times. Resultant from the multiresolution decomposition of each one of the temporary decimated patterns, there were generated patterns consisting the wavelet coefficients corresponding to the approximation of level six and the details of levels one to six. Approximation of level six was always conserved, because by frequency considerations it was assumed that the in that level of scale, the patterns contained information on the presence or not of the ERP. Considering each detail and approximation has half the samples of the approximation in the previous level, starting with a signal of 256 samples, is arrived, the sixth approach has only four samples. In table III the amount of samples of the approximation in the sixth level and of each one of the details in different levels is detailed for the decimated patterns of 256 samples.

Table III. Amount of samples for different decomposition levels

Decomposition depth		Number of samples
Level 1	Detail	128
Level 2	Detail	64
Level 3	Detail	32
Level 4	Detail	16
Level 5	Detail	8
Level 6	Detail	4
	Approximation	4

From the decimated single trial patterns the decomposition was made with different wavelet families, obtaining in this form the wavelet coefficients corresponding to the approximation in the sixth level and the details of levels one to six. Then the coefficients corresponding to the sixth level approximation and to the details of different levels were taken, to form with these the patterns that were presented to the classifier.

The table IV presents the performance of the single layer perceptron in the detection of patterns corresponding to decompositions of the patterns with some representative mother wavelets of, taking in each case different amount of details.

Table IV. System performance for different wavelets and details coefficients combinations

	Wavelet: Biortogonal 3.9			**Wavelet**: Daubechies 9		
Details	Accuracy [%]	Sensitivity	Specificity	Accuracy [%]	Sensitivity	Specificity
1-6	61.00	0.68	0.54	63.25	0.71	0.55
2-6	61.50	0.71	0.52	63.75	0.72	0.55
3-6	59.50	0.69	0.5	65.25	0.70	0.60
4-6	63.75	0.75	0.53	66.25	0.72	0.60
5-6	65.00	0.77	0.53	69.25	0.74	0.64
	Wavelet: Coiflets 4			**Wavelet**: Symlets 8		
Details	Accuracy [%]	Sensitivity	Specificity	Accuracy [%]	Sensitivity	Specificity
1-6	61.00	0.70	0.52	63.00	0.73	0.53
2-6	64.25	0.75	0.54	64.5	0.73	0.56
3-6	66.75	0.75	0.58	62.50	0.72	0.53
4-6	64.00	0.76	0.52	65.75	0.76	0.55
5-6	64.75	0.75	0.55	60.50	0.72	0.49

4. Conclusions

From the analysis of results it can be seen that the best performance is superior when using Daubechies 9 as mother wavelet for the decomposition, this can be seen in general for every mother wavelet belonging to this family that were used. It can also be noticed that the system performance is lower when performing the decomposition with some wavelet families than using a temporal base. It is also important to highlight that the sensibility was in general bigger than the specificity, result that in case of being consistent must be taken into account on the brain computer interface design with the objective of improving its efficiency.

A similar analysis was also made using averages of 2, 3, 4, 5, 6, 7, 8, 9 and 10 trials reaching accuracy values of about 75% with averages of ten trials applying DDWT and almost of 70% presenting to the SLP the ten trial averages of temporal registers.

In spite of being an initial step in this line of work encouraging results have been achieved, since using a simple and probably not optimal classifier and using registers of only one channel with averages of two trials acceptable classification rates were achieved.

However, to achieve a practically exploitable development the performance of the system should be improved; to achieve this goal some possible directions to consider are the inclusion of information obtained from other channels and the experimentation with other classifiers. Another possibility consists in designing a method to estimate an optimal discriminative dictionary.

Acknowledgements
This work was supported by the Universidad Nacional de Entre Ríos (with PID 6101 and 6106), the Universidad Nacional de Litoral (with CAI+D 012-72), STIC-AMSUD program (09STIC01), the Agencia Nacional de Promoción Científica y Tecnológica (with PAE 37122, PAE-PICT-2007-00052) and the Consejo Nacional de Investigaciones Científicas y Técnicas (CONICET) from Argentina.
.

References

[1] R. Acevedo, C. Bonell, G. Gentiletti, J. Aldonate, C. Mercuri, S. Escobar and J. Biurrun 2006. Potenciales evocados auditivos de tronco cerebral: Aspectos tecnológicos del registro *Revista Argentina de Bioingeniería* (SABI), Vol. 12 No. 1, pp. 10-17.
[2] G. Gentiletti, C. Tabernig and R. Acevedo 2007. Interfaz cerebro-computadora: Estado del arte y desarrollo en Argentina *Revista Argentina de Bioingeniería* (SABI), Vol. 13 No. 1, pp. 13-19, ISSN: 0329-5257
[3] L. A. Farrel and E. Donchin 1988. Taking off the top of your head: toward a methal prothesis utilizing event related brain potentials *Electroencephalography and Clinical Neurophysiology*, vol. 70, pp. 510-523.
[4] R. Wolpaw, N. Birbaumer, D. J. McFarland, G. Pfurtscheller, and T. M. Vaughan 2002. Brain-computer interfaces for communication and control *Clinical Neurophysiology*, vol. 113, pp.767-791.
[5] J. del R. Millán, P. W. Ferrez, F. Galán and E. Lew 2007. Non-invasive brain-machine interaction *International Journal of Pattern Recognition and Artificial Intelligence*, vol. 20, pp. 1-13.
[6] http://www.bci2000.org/BCI2000/Home.html
[7] G.G. Gentiletti Faenze, J.G. Gebhart, M. Richard, E. García Cossio, R.C. Acevedo and V. Medina Bañuelos. Robotics Simulation Platform for BCI Application: A Wheelchair Driving Example Using P300 Paradigm *Proceedings of the 4th International Brain-Computer Interface Workshop and Training Course*, Graz (Austria), 2008
[8] Chen, Donoho and Saunders, "Atomic Decomposition by Basis Pursuit", SIAM REVIEW, Vol. 43,No . 1, pp . 129–159, 2001.
[9] H.L. Rufiner 2009, "*Análisis y modelado digital de la voz: técnicas recientes y aplicaciones*", Ed. UNL, Colección Ciencia y Técnica.

Journal of Physics: Conference Series **313** (2011) 012012 doi:10.1088/1742-6596/313/1/012012

Short-time multifractal analysis: application to biological signals

Roberto F. Leonarduzzi[1], Gastón Schlotthauer[1] and María E. Torres[1,2]

[1]Lab. Signals and Nonlinear Dynamics, Faculty of Engineering,
Universidad Nacional de Entre Ríos, Argentina

Abstract. Some signals obtained from biological systems evince a great complexity. Recently, new tools which allow the extraction of information from them have been proposed. In particular, multifractal analysis gives a quantification of the degree and distribution of irregularities in a signal. A possible approach for this analysis is the one based on wavelet leaders. In this work, the use of wavelet leader based multifractal analysis in short-time windows is proposed in order to analyze the evolution of the multifractal behavior of biological signals. In particular, applications of this technique to the detection of ischemic episodes in heart rate variability signals and to voice activity detection are examined. It is shown that the study of the time evolution of indexes obtained with the proposed new method gives useful information hidden in the signals.

1. Introduction

Some biological systems show a great complexity. Signals from these systems are generally nonlinear and nonstationary, and show complex and irregular fluctuations even under resting conditions. This fact defies the traditional concept of *homeostasis*, according to which healthy biological systems are in equilibrium and self-regulated to reduce variability [1]. On the contrary, these systems show erratic fluctuations resembling those of systems operating far from equilibrium. Moreover, these fluctuations contain hidden information about the control systems operating in this non equilibrium state [2].

In the 1990s, tools from statistical physics were first used to cope with this complexity in order to extract this hidden information coded in the fluctuations of time series (TS). In particular, the multifractal formalism (MFF) allows to study scaling phenomena and long-range correlations in TS, providing a quantification of its singularities distribution. A statistical approach based on the continuous wavelet transform gave rise to a unified multifractal description [3]. Variants were proposed, based on the wavelet transform modulus maxima (WTMM) [4], the discrete wavelet transform (DWT) [5] and, more recently, the wavelet leaders (WLs) [6],[7].

Multifractal analysis has been successfully applied to the analysis of numerous biological signals: DNA sequences [4], heart rate variability [5], EEG [9], human gait [10] and speech [11], amongst others.

In the present work, we propose the application of WLs based multifractal analysis in short-time windows in order to analyze the evolution of the multifractal behavior of the signal. The selection of the window length is discussed using artificial signals. Two applications of this technique to real

[2] Author to whom any correspondence should be addressed. E-mail: metorres@santafe-conicet.gov.ar

biological signals are presented: detection of ischemic episodes in HRV signals and voice activity detection (VAD.)

2. Materials and methods

In this section a brief review of the tools used in the present work is given. The definitions of Hölder exponent and the singularity spectrum are presented. The WLs and the corresponding multifractal formalism are described, following [7]. Finally the short-time multifractal analysis proposed in the present work is presented. The signals used for the experiment are described in the last subsection.

2.1. Hölder exponent and singularity spectrum

Given a point $t_0 \in \mathbb{R}$ and a real constant $\alpha \geq 0$, a function $f : \mathbb{R} \to \mathbb{R}$ is said to be $C^\alpha(t_0)$ if there exists a constant K > 0 and a polynomial P_{t_0} of degree less than α, such that $|f(t) - P_{t_0}(t)| \leq C|t - t_0|^\alpha$. The Hölder exponent $h_f(t_0)$ of f in t_0 is defined as $h_f(t_0) = \sup\{\alpha : f \in C^\alpha(t_0)\}$ [6]. It measures the local regularity of f in t_0. Small (close to 0) values of the Hölder exponent denote strong and sharp singularities, whereas large values denote smooth ones.

When analyzing a singular almost everywhere signal, it might be useful to know the distribution of the singularities with a given Hölder exponent. This information is provided by the singularity (or multifractal) spectrum (SS): it measures the *amount* of singularities with a given Hölder exponent. Formally, it is defined as the Hausdorff dimension of the set of points whose Hölder exponent is h: $D(h) = \dim_{\mathrm{H}}\{t \in \mathbb{R} : h_f(t) = h\}$.

2.2. Multifractal formalism

The determination of the SS of a signal is important to analyze the singularities present in a given signal. However, it can not be numerically determined from its definition because of the limitations imposed by finite resolution and the sampling of signals [7]. In order to cope with this situation, a *multifractal formalism* (MFF) was introduced. The MFF provides an alternative way to compute the SS based on easily computable quantities: the structure functions (SFs). As already mentioned, a new MFF, based on WLs, was recently proposed [7]. This approach solves many of the drawbacks of previous methods. Its main features are the fact of being based on the DWT, which allows an efficient implementation in terms of filter banks, and its numerical stability.

2.3. Wavelet leaders

Let ψ_0 be a mother wavelet with compact support and a number $N \geq 1$ of vanishing moments. Let $\{\psi_{j,k}(t) = 2^{-j}\psi_0(2^{-j}t - k), j \in \mathbb{N}, k \in \mathbb{N}\}$ be the orthonormal basis of $L^2(\mathbb{R})$ formed by templates of ψ_0 dilated by scales 2^j and translated to time positions 2^jk. The DWT coefficients of f are obtained by means of the inner products: $d_f(j,k) = \int_{\mathbb{R}} f(t) 2^{-j}\psi_0(2^{-j}t - k)\mathrm{d}t$.

A special notation is introduced for dyadic intervals. Let $\lambda = \lambda_{j,k} = [k2^j, (k+1)2^j)$ so that $d_\lambda \equiv d_f(j,k)$. Finally, let 3λ be the union of the dyadic interval λ and its two adjacent intervals: $3\lambda_{j,k} = \lambda_{j,k-1} \cup \lambda_{j,k} \cup \lambda_{j,k+1}$.

The WLs are defined as:

$$L_f(j,k) \equiv L_\lambda = \sup_{\lambda' \subset 3\lambda} |d_{\lambda'}|. \tag{1}$$

In other words, the computation of the WLs for a given time and scale involves the search of the greatest wavelet coefficient in a narrow time neighborhood, for all finer scales.

2.4. Wavelet leader based multifractal formalism

SFs are computed from WLs:

$$S_L(q,j) = \frac{1}{n_j}\sum_{k=1}^{n_j}\left|L_f(j,k)\right|^q. \quad (2)$$

If the signal f shows some form of self-similarity, SFs decay as power laws of the scales. The exponents of these power laws are called *scaling exponents* (SEs):

$$\varsigma_L(q) = \liminf_{j\to 0}\left(\frac{\log_2 S_L(q,j)}{j}\right). \quad (3)$$

In practice, SEs are obtained as the slopes of the linear regressions of SFs *vs.* scales, in a logarithmic graph.

Finally, the SS can be obtained from the SEs via a Legendre transform (LT):

$$D(h) = \inf_{q\neq 0}\left(1 + qh - \varsigma_L(q)\right). \quad (4)$$

2.5. Short-time multifractal analysis

MFA of a signal provides global information concerning its singularities distribution. However, in some cases local information is required, e.g. in the detection of transient states that cause changes in the singular behavior of signals. With this in mind, in the present work we propose the application of MFA in short-time windows to analyze the evolution of the multifractal behavior of signals. Avoiding the study of the whole SS, we propose to use multifractal indexes to summarize the information in a few coefficients and to analyze their evolutions.

As first index we propose to use the dominant regularity h_{max}, which is defined as the Hölder exponent for which the spectrum has its maximum: $D(h_{\max}) > D(h), \forall h \neq h_{\max}$. Small (close to zero) values of this index denote signals with predominantly sharp singularities, whereas large values correspond to signals that mainly have smooth singularities.

As second index we consider the spectrum's width at 80% of its height, W_{80}. It measures the *multifractality* of the signal. Given $h_1 < h_{\max} < h_2$ such that $D(h_1) = 0{,}8\, D(h_{\max})$, $h_1 < h_{\max}$, and $D(h_2) = 0{,}8\, D(h_{\max})$, $h_2 > h_{\max}$, then $W_{80} = h_2 - h_1$. This last index measures the local regularity distribution. Large values reflect that the signal has singularities of very different Hölder exponent (wide spectrum). Small values reflect signals with singularities that have similar Hölder exponents (narrow spectrum). When all the singularities of a signal have the same Hölder exponent, the SEs is a line with slope equal to this Hölder exponent and the SS is a point. These kinds of signals are called monofractals.

With these two indexes, we propose to perform a short-time multifractal analysis, meaning to compute each index in a sliding window applied to the signal. The values obtained in each window are assigned to its middle point.

2.6. Signals and databases

We have considered a synthetic signal and two types of real signals: RR time series and speech signals.

In order to analyze the influence of the window length, artificial signals formed by the concatenation of three segments of fractional Brownian motion (fBm) with different Hurst parameters were used: $s=(s_1|s_2|s_3)$. fBm is monofractal: it is singular in almost every point and all singularities have the same Hölder exponent, which is equal to the Hurst parameter [7]. Segments s_1 and s_3 consisted in 2^{18} samples of fBm with the same Hurst parameter H_1. Segment s_2 consisted in 2^{19} samples of fBm with Hurst parameter H_2 (figure 1.a.) Different combinations of H_1 and H_2 and 100 realizations of each combination were analyzed.

For the analysis of HRV, the European ST-T database (ESDB) [12] was used. It comprises two-channel ECG records, two hours long each, of patients suffering from myocardial ischemia. Each record contains at least one ischemic episode. Records are digitized with a sampling frequency of 250 Hz. For each record, the ESDB has one annotations file with information provided by experts. It includes the position of each beat, the beginning and end of each ischemic episode (obtained from the deviation of the ST segment and T wave of the ECG) and the characteristics of each beat (normal, ectopic, noisy, etc).

From the annotations file, RR interval series were obtained considering the difference between two consecutive beats. All intervals corresponding to abnormal beats were eliminated. This procedure prevents the inclusion of artifacts, avoiding in this way the need of a filtering stage. Finally, the signals were uniformly interpolated with cubic splines and a sampling frequency of 4 Hz, given that an uniformly sampled signal is required for the DWT computation.

For the study of voice activity detection, the database BEPPA [13] was used. This database comprises a set of words and sentences of argentinean Spanish. Records are digitalized with a sampling frequency of 48 kHz. In this preliminary work, only the subset of monosyllabic words MW02 was used. This list comprises 10 monosyllabic words.

3. Results

3.1. Influence of window´s length

In order to study the influence of the window´s length, we consider the synthetic signals described in Sec. 2.6. Given that the signal to be analyzed is a concatenation of monofractals, almost constant values of h_{max}, close to the corresponding Hurst parameter, and small values of W_{80} are expected for each of the time segments corresponding to the concatenated signals.

Figure 1 shows one realization, with H_1=0.3 and H_2=0.7 (a), and the evolutions of the multifractal indexes for two window lengths: 2^{15} samples (b-c) and 2^{11} samples (d-e).

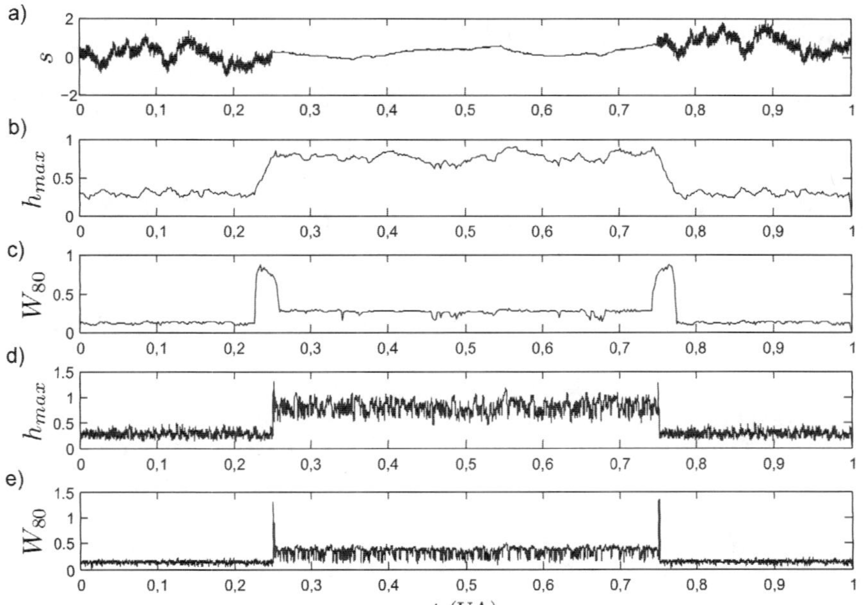

Figure 1. One realization of the artificial signals described in section 2.6, with H_1=0.3 and H_2=0.7 (a) and time evolutions of h_{max} and W_{80} for window´s lengths of 2^{15} samples (b-c) and 2^{11} samples (d-e).

It can be appreciated that when the window length is large (2^{15} samples), h_{max} assumes values close to the expected value. On the contrary, when a small window (2^{11} samples) is used, the evolution shows wider fluctuations around it. A similar phenomenon can be observed in the evolutions of W_{80}. This loss of precision when a small window is used is due to the fact that the number of samples limits the number of scales that can be obtained to perform the DWT. On one hand, this causes a loss of precision in the estimation of the WLs, which requires the search of the largest wavelet coefficient in a wide range of scales. On the other hand, the range of scales over which the linear regressions are performed to obtain the SEs from the SFs is also reduced, causing an additional loss of precision.

The means and standard deviations (*std*), for both window's lengths considered and 100 realizations of segments s_1 and s_2, are shown in figure 2 (a) and (b), respectively. Points corresponding to windows of 2^{15} and 2^{11} samples are shown in black and gray, respectively. Dots and crosses are used for segments with Hurst parameter H_1 and H_2, respectively. It can be seen that the mean of h_{max} is close to the Hurst parameter of the segment for both window sizes, while the *std* is larger when using the short windows. Similar results have been obtained in a more extensive study performed by the authors, using different values of H_1 and H_2 [14]. These experiments illustrate the loss of precision in the estimation of the multifractal indexes when using short windows.

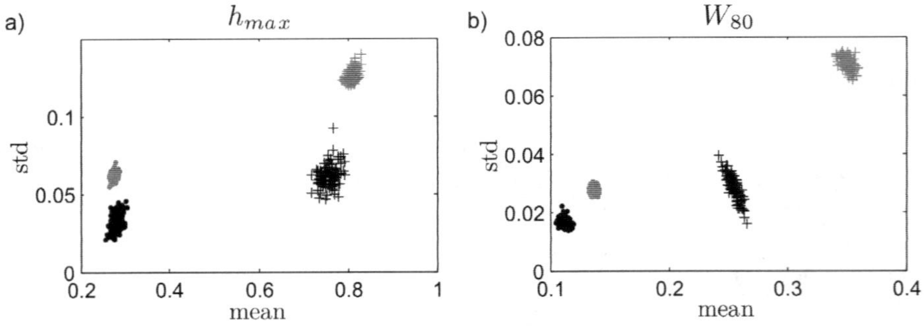

Figure 2. Mean *vs.* *std* of indexes h_{max} (a) and W_{80} (b) for 100 realizations of the artificial signals described in the text. Dots: segments with Hurst parameter H_1. Crosses: segments with Hurst parameter H_2. Black: 2^{15} samples long window. Gray: 2^{11} samples long window.

The use of short windows allows for a better temporal resolution. This can be seen clearly in figure 1 (c) and (e), for the index W_{80}. The transitions between segments cause A_{80} to show large values on a narrow region when the narrow window is used (c,) and on a wide one when the large window is used (e).

These experiments suggest that the window length is a parameter that should be chosen carefully and as a tradeoff between temporal resolution and accuracy in the estimation of the SS.

3.2. Analysis of ischemic episodes

It was shown in [8] that multifractal analysis of HRV conveys useful information concerning the dynamic changes that take place during myocardial ischemia. The SEs and SS corresponding to a normal zone and an ischemic zone of record e0154 of the ESDB are shown in figure 3 (a) and (b), respectively. It can be seen that, in the presence of ischemia, the SS becomes wider and the predominant regularity h_{max} moves to higher values (singularities become smoother). This information can also be seen in the SEs. Indeed, the SEs corresponding to the ischemic zone shows a greater curvature and its least square approximation line shows a larger slope (smoother predominant local regularity) than those of the normal zone.

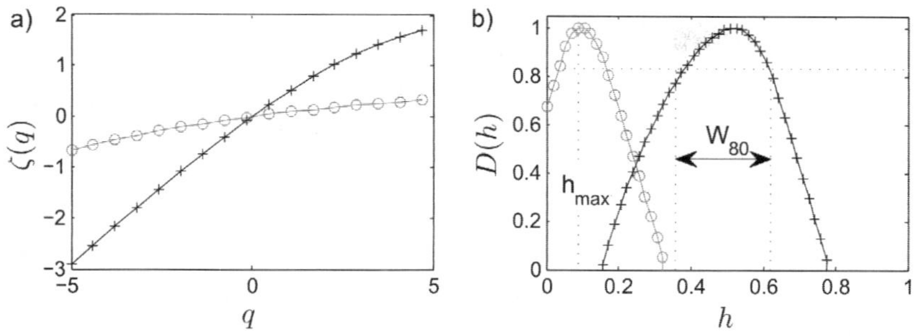

Figure 3. Scaling exponents $\varsigma_L(q)$ (a) and singularity spectrum $D(h)$ (b) corresponding to a normal zone (gray circles) and an ischemic zone (black crosses). Also, the indexes h_{max} and W_{80} are illustrated.

In the short-time analysis of RR signals, 2^{11} samples (512 s) long windows and 128 samples time shifts were used. Figure 4 shows a RR signal, with an ischemic episode indicated by vertical gray lines (a), and the time evolutions of h_{max} (b) and W_{80} (c). It can be appreciated that, when approaching the ischemic zone, both indices show a sharp rise and that after the episode they move down to values similar to those obtained before the ischemic episode. While analyzing these evolutions, it must be taken into consideration that myocardial ischemia entails a series of alterations in which metabolic changes precede hemodynamic changes, which in turn precede electrocardiographic changes. As a consequence, the ischemic episode is reflected in the RR signal beyond the bounds indicated by the marks obtained from the ECG.

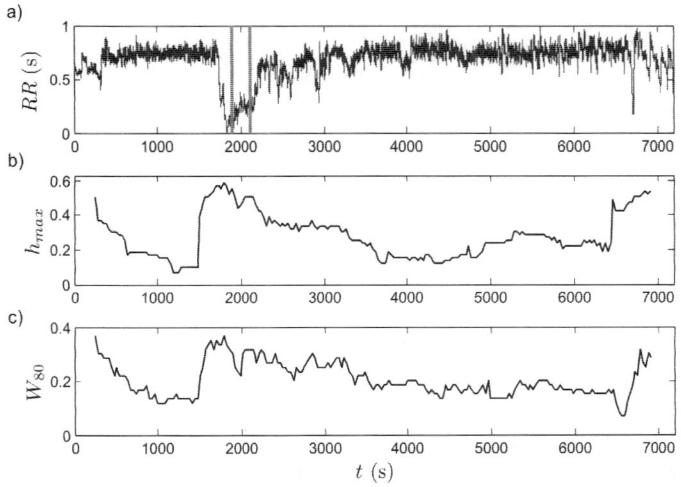

Figure 4. RR signal (a) with an ischemic episode indicated by vertical gray lines and the windowed time evolutions of h_{max} (b) and W_{80} (c).

3.3. Voice activity detection

A fractal analysis of speech signals was proposed in [11] with the objective of obtaining useful information for voice activity detection and phoneme recognition applications. This analysis was justified by turbulence phenomena in the air flow during speech production and other nonlinear dynamics of the human vocal-tract system. This work was based on the concept of fractal dimension, which characterizes the complexity and irregularity of a signal using, unlike multifractal analysis, only one index.

This approach is extended in the present work to a multifractal analysis applied to VAD. The hypothesis that the regularity of voice signals is different than the one corresponding to the background noise is assumed.

A short-time multifractal analysis was performed, using 2^{11} samples (43 s) long windows and 256 samples time shifts.

As a comparison, the time evolution of the zero-crossing rate (*ZC*), a standard index in VAD, was computed. The same window size and time shift than in the multifractal approach were used.

Figure 5 shows the sonogram of the Spanish word "flan" [flán] (a) and the time evolutions of the multifractal indexes and ZC (b-d). The evolutions of the multifractal indexes are shown in semi-log graphs to highlight structures that take place when the indexes' values are small. It can be seen that h_{max} has a similar behavior than *ZC*, whereas W_{80} fails to properly detect the final portion of the word. If the interval between 0.3 s and 0.5 s, corresponding to the non-sonorant fricative phoneme /f/, is carefully examined, it can be observed that in figure 5 (d) *ZC* fails to recognize this phoneme whereas figure 5 (b-c) indicate that both multifractal indexes display a small rise in this region . On the other hand, both h_{max} and W_{80} show a small peak around 0.3 s (indicated by the circles and arrows in figure 5, (b-c)), which is coincident with the beginning of the phoneme. During all the phoneme's length, W_{80} remains with a value slightly larger than the one corresponding to the noise that preceeds it..

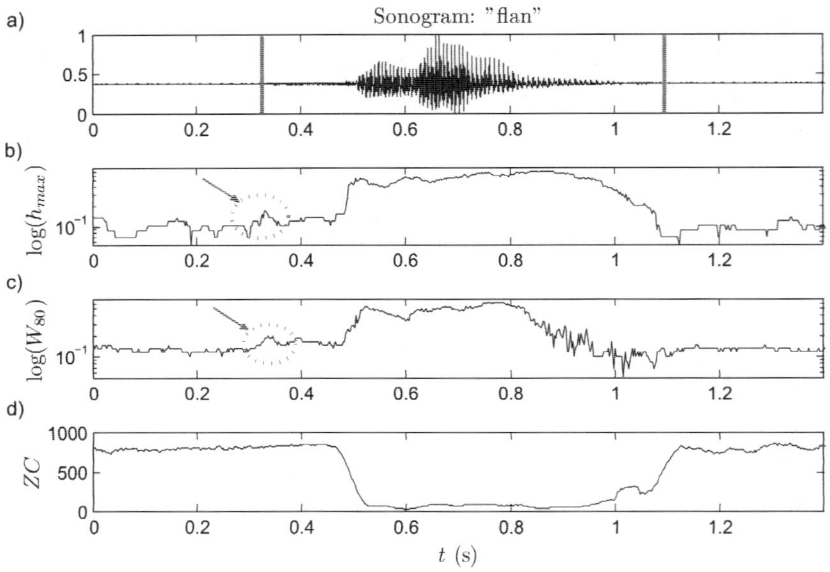

Figure 5. Sonogram of the Spanish word "flan" (a), with a manual segmentation performed by the authors indicated by vertical gray lines, and windowed time evolutions of indexes h_{max}, W_{80} (in a semi-log graph) and the zero-crossing rate *ZC* (b, c and d, respectively).

These experiments suggest that the joint use of the proposed multifractal indexes might provide proper voice activity detection, even for the fricative phonemes where the zero-crossing rate is similar to the one of the background noise. This is an observation that should be taken into consideration in VAD in Spanish language, given the importance of this type of phonemes in this language.

4. Conclusions

In the present work we have proposed to use multifractal analysis based in wavelet leaders, in short-time sliding windows to analyze biological signals. The application of this technique to two different kinds of real biological signals was examined. Results suggest that the study of the windowed time evolution of indexes derived from the SS is useful for the detection of transient alterations that cause changes in the regularity of the signal.

The effect of the window´s lenght has been discussed with synthetic signals. . This parameter should always be chosen carefully as a tradeoff between temporal resolution and accuracy in the estimation of the SS.

References

[1] Goldberger A, Amaral L, Hausdorff J, Ivanov P, Peng C and Stanley H 2002 Fractal dynamics in physiology: alterations with disease and aging *Proc. National Academy of Sciences* vol 99 pp 2466-72

[2] Stanley H, Amaral L, Goldberger A, Havlin S, Ivanov P and Peng C 1999 Statistical physics and physiology: monofractal and multifractal approaches *Physica A* **270** 309-324

[3] Arneodo A, Bacry E and Muzy J 1995 The thermodynamics of fractals revisited with wavelets *Physica A* **213** 232-275

[4] Arneodo A, Audit B, Decoster N, Muzy J and Vaillant C 2002 Wavelet based multifractal formalism: application to DNA sequences, satellite images of the cloud structure and stock market data. *The Science of Disasters: Climate Disruptions, Heart Attacks, and Market Crashes* ed A Bunde *et al.* (Berlin: Springer Verlag) pp 26-102

[5] Veitch D and Abry P 1999 A wavelet-based joint estimator of the parameters of long-range dependence *IEEE Transactions on Information Theory* **45** 878-97

[6] Jaffard S, Lashermes B and Abry P 2006 Wavelet leaders in multifractal analysis *Wavelet analysis and applications* ed T Qian *et al.* (Cambridge, MA: Birkhäuser) pp. 219-64

[7] Lashermes B, Jaffard S and Abry P 2005 Wavelet leader based multifractal analysis *Proc. IEEE International Conference on Acoustics Speech and Signal Processing* vol 4 pp 161-64

[8] Godoy J L, Aniño M M and Torres M E 2003 Análisis multifractal de la regulación autonómica del ritmo cardíaco en episodios isquémicos, *Revista Argentina de Bioingeniería* **9** 24-9

[9] Popivanov D, Stomonyakov V, Minchev Z, Jivkova S, Dojnov P, Jivkov S Christova E and Kosev S 2006 Multifractality of decomposed EEG during imaginary and real visual-motor tracking *Biological Cybernetics* **94** 149-56

[10] West B and Scaffeta N 2005 A multifractal dynamical model of human gait *Fractals in biology and medicine* vol 4 ed G Losa *et al.* (Basel: Birkhäuser) pp 131-40

[11] Maragos P and Potaminos A 1999 Fractal dimensions of speech sounds: computation and application to automatic speech recognition *Journal of the Acoustical Society of America* **105** 1925-32

[12] Taddei A *et al.* 1990 The European ST-T Database: development, distribution and use *Computers in cardiology* pp 177-80

[13] Aronson L, Estienne P, Milone D, Martínez C, Rufiner H y Torres M E 2007 Base de Datos Batería de Evaluación para Pacientes con Prótesis Auditivas (BEPPA) *Revista de la Federación Argentina de Sociedades de Otorrinolaringología* **14** 17 – 24

[14] Leonarduzzi R F, Torres M E and Schlotthauer G 2010 Short time multifractal analysis: evaluation of window size and application to biological signals *Physica A* (in preparation)

[15] Gamero L, Vila J and Palacios F 2002 Wavelet transform analysis of heart rate variability during myocardial ischemia *Medical and Biological Engineering and Computing* **40** 72-8

Biomechanics of Ergometric Stress Test: regional and local effects on elastic, transitional and muscular human arteries

G Valls[1], J Torrado[1], I Farro[1], D Bia[1], Y Zócalo[1], S Lluberas[1], D Craiem[2], RL Armentano[1, 2]

[1]Departamento de Fisiología, Facultad de Medicina, Universidad de la República, Montevideo, Uruguay.
[2] Facultad de Ingeniería y Ciencias Exactas y Naturales, Universidad Favaloro, Buenos Aires, Argentina.

dbia@fmed.edu.uy

Abstract. Ergometric exercise stress tests (EST) give important information about the cardiovascular (CV) response to increased demands. The expected EST-related changes in variables like blood pressure and heart rate are known, but those in the arterial biomechanics are controversial and incompletely characterized. In this context, this work aims were to characterize the regional and local arterial biomechanical behaviour in response to EST; to evaluate its temporal profile in the post-EST recovery phase; and to compare the biomechanical response of different to EST. Methods: In 16 non-trained healthy young subjects the carotid-femoral pulse wave velocity and the carotid, femoral and brachial arterial distensibility were non-invasively evaluated before (Rest) and after EST. Main results: The EST resulted in an early increase in the arterial stiffness, evidenced by both, regional and local parameters (pulse wave velocity increase and distensibility reduction). When analyzing conjunctly the different post-EST recovery stages there were quali-quantitative differences among the arterial local stiffness response to EST. The biomechanical changes could not be explained only by blood pressure variations.

1. Introduction

It is known that when evaluating the cardiovascular (CV) system studies performed at rest and during conditions of increased requirements (i.e. ergometric exercise stress test, EST) give complementary information. The expected changes and temporal profiles of variables like heart rate and blood pressure (BP) in response to EST have been characterized, and altered responses of such variables have shown to be associated with CV risk and/or prognostic. Then, the evaluation of changes in HR and arterial pressure in response to EST is widely used in clinical practice in the evaluation of functional capability, in CV diagnosis and patients follow up. On the other hand, it is recognized the role of the arterial biomechanics in the CV physiology and disease development.

Taking into account this, recent international guidelines [1] included the study of basal vascular biomechanics in the CV risk stratification, diagnostic, prognostic and follow up. However, it is noteworthy that subjects with normal basal vascular biomechanics could have an abnormal response to increased CV demands. This has been frequently overlooked, and in our knowledge only few studies evaluated the vascular biomechanics during the exercise. Furthermore, the studies results are controversial.

Taking into account the structural and functional differences among arteries and their different role in the cardiovascular system, the biomechanical response to exercise could differ, depending on the exercise developed, as well as on the vascular region and vessels analyzed (i.e. elastic arteries, like the carotid vs. muscular arteries like the femoral) [2], [3]. Hence, an adequate characterization of the vascular response to a particular EST requires evaluating regional parameters, and a simultaneous analysis of different arteries local response.

Several approaches are used to evaluate, non-invasively, the arterial system biomechanical behaviour. Among them, the analysis of the pressure-diameter relationship in a given arterial segment is considered the gold standard in 'local' stiffness evaluation, while the pulse wave velocity measurement is accepted as the gold standard for assessing 'regional' stiffness [1].

In this context, this work aims were: a) to characterize regional (carotid-femoral) and local (elastic, muscular and transitional arteries) changes in vascular stiffness in response to EST, b) to determine the temporal profile of the vascular biomechanics changes during the post-exercise recovery phase, and c) to compare the biomechanical response to exercise among different arterial segments, and d) to analyze the relationship between the biomechanical changes and BP, taking into account the pressure dependence of the arterial stiffness (isobaric analysis). In other words, the work general aim was to determine the arterial stiffness changes associated to the EST using complementary approaches and gold standard parameters.

2. Materials and methods

2.1 Subjects

Sixteen healthy, non-trained, active adults (11 women), without CV risk factors and free of CV and other chronic diseases were studied. None of the subjects was taking CV acting medications. The study was carried out according to international ethic rules and the Helsinki Declaration principles. All procedures were approved by the Republic University (Uruguay) Human Research Committee. Subjects gave written informed consent. The subjects' main characteristics are detailed in the Table.

Table. Subjects main characteristics.

Age [Years]	21	±	1
Body weight [Kg.]	63	±	8
Body height [m]	1.65	±	0.15
BMI [Kg/m^2]	22.1	±	2.0
Hip perimeter [cm]	94.38	±	7.47
Waist perimeter [cm]	73.33	±	5.64

Values expressed as mean ± standard deviation

2.2 Experimental protocol

In each experimental session, brachial BP, electrocardiogram and regional (left carotid-femoral pulse wave velocity, PWV) and local (distensibility) arterial stiffness measurements were obtained before (Rest) and after bicycle EST. Post-exercise recordings were obtained 0-1 (PE1), 4-5 (PE2) and 9-10 (PE3) minutes after the exercise. All measurements were done with the subject in supine position, in a quite room, with controlled-temperature (22°C) [1].

2.3 Arterial non-invasive studies: pulse wave velocity and arterial distensibility

Arterial non-invasive studies were done according to international consensus [1]. To evaluate the aorto-iliac regional biomechanical behaviour, the carotid-femoral PWV was measured using strain gauge mechano-transducers (Motorola MPX 2050, Motorola Inc., Corporate 1303 E.Algonquin Road, Schaumburg, Illinois 60196, USA) placed simultaneously on the skin over the carotid and femoral arteries [4]. Signals were recorded and digitized for off-line analysis.

Given the distance (Δx) between the mechano-transducers, the PWV was calculated for 10 consecutive pulses as $PWV=\Delta x/\Delta t$, where Δt is the time difference between the maximum upstroke dP/dt of the carotid and femoral waves (Figure 1). The average PWV was used for further analysis. All measurements were made by the same operator. PWV variation coefficient was less than 5%.

The carotid, brachial and femoral distensibility was characterized through the analysis of the systo-diastolic diameter-pressure relationship. To evaluate the diameter each artery was visualized longitudinally by high resolution B-Mode ultrasound (Sampling rate: 30 Hz; 7.5-MHz probe; Portable Ultrasound System, Model: Aloka SSD210, ALOKA CO., LTD. Tokyo, Japan) and video sequences (10-15s) were recorded [4]. Then, the recorded sequences were analyzed off-line using an automated step-by-step algorithm applied to each digitalized image that allowed obtaining the diameter signal [5].

To determine the carotid and femoral peak systolic (SBP) and minimum diastolic (DBP) pressure levels, taking into account that pulse pressure (PP) may differ between central and peripheral arteries, the carotid and femoral pressure waves were calibrated using brachial pressure values obtained by sphygmomanometry [1]. To do this, since DBP and mean pressure (MBP: MBP=DBP+ (SBP-DBP)/3) do not vary along the arterial system (especially in supine position) the DBP and MBP values measured in the brachial artery were assigned, respectively, to the minimum and average values of the carotid and femoral pressure waves. Then, the carotid and femoral SBP and PP levels were obtained [1] [2].

Subsequently, the effective arterial compliance (ADE), a parameter commonly used in clinical practice, was calculated: ADE = (SD-DD / SBP-DBP) /DD, where SD and DD are the systolic and diastolic internal diameters, respectively.

Finally, in order to discriminate potential pressure-dependent and pressure-independent (i.e. smooth muscle tone-related) biomechanical changes associated with EST, the passive isobaric arterial distensibility (ADPI) was calculated for the pressure levels observed after EST. To this end, the pressure-diameter logarithmic relationship was obtained at rest and the ADPI quantified for the pressure levels observed in response to the exercise. Then, ADPI would represent the distensibility expected at rest, if pressure levels were those observed in PE1, PE2 and PE3. The ADPI and ADE comparison allowed evaluating if the arterial distensibility variations could be explained by the pressure changes [6].

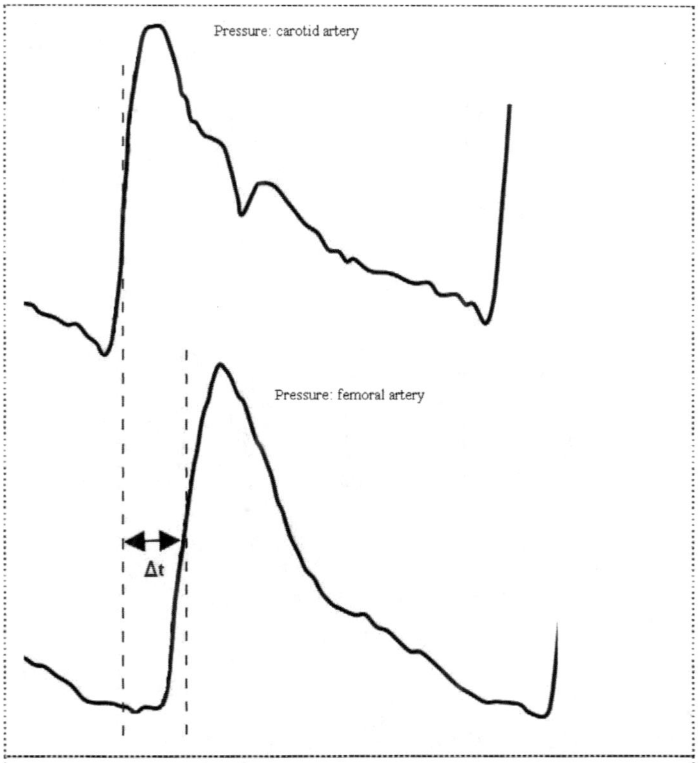

Figure 1. Mechanographic pressure wave recordings

2.4 Statistics

Data are shown as mean value +/- standard deviation.

Changes in pressure, heart rate (HR), PWV, ADE, ADPI and arterial diameter, associated with the EST were evaluated using ANOVA + Bonferroni tests. Differences between ADE and ADPI were evaluated using two tailed paired Student t test. A $P<0.05$ indicated significant statistical differences.

3. Results

Figure 2 shows brachial BP (top), HR (middle) and mean arterial diameter (bottom) levels, obtained at rest and at different stages in the EST recovery phase (PE1, PE2 and PE3). Note that PE1showed differences in both, systolic and diastolic pressure levels, with respect to rest conditions, PE2 and PE3 stages. However, as was expected, as a result of the systolic (increase) and diastolic (decrease) changes in pressure, MBP kept unchanged after the EST.

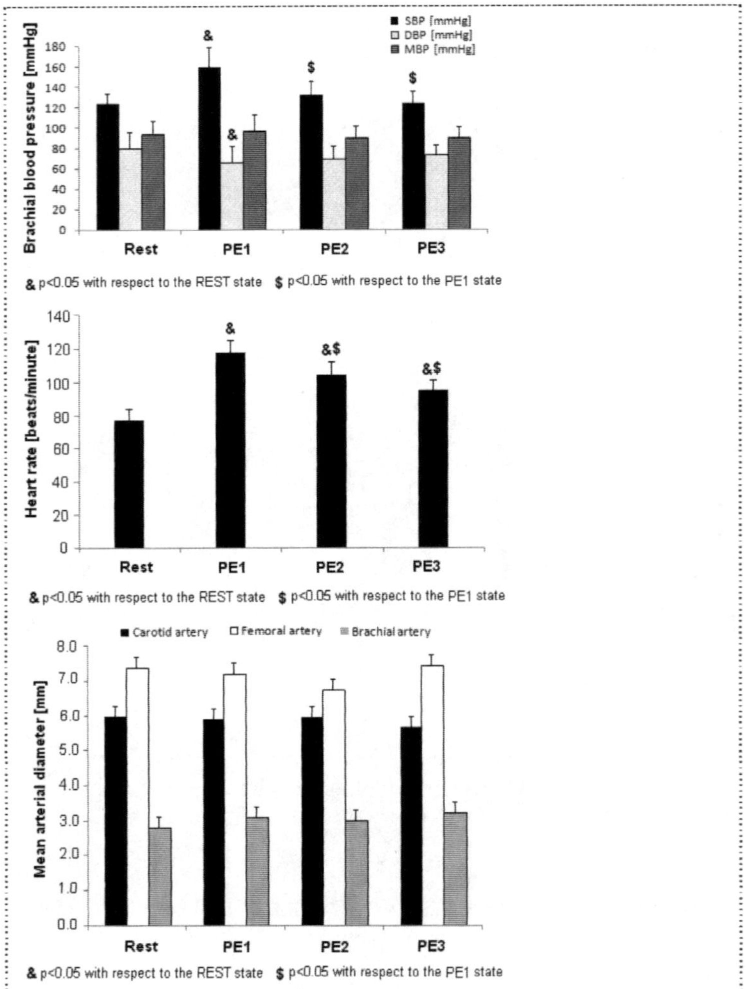

Fig 2: Brachial pressure (Top), heart rate (Middle) and arterial diameter (Bottom), obtained at rest and after exercise (PE1-PE3).

Figure 3 shows ADE and ADPI for the elastic (carotid), muscular (femoral) and transitional (brachial) arteries, at rest, and after EST. It is noteworthy, that in all the studied arteries, EST resulted in an early decrease in DAE, with quantitative differences among the vascular segments studied. Furthermore, after PE1, DAE changes varied, depending on the analyzed artery: in the femoral, DAE increased, reaching levels higher than those found at rest; in the carotid, DAE returned to basal values and it kept in low levels in the brachial artery.

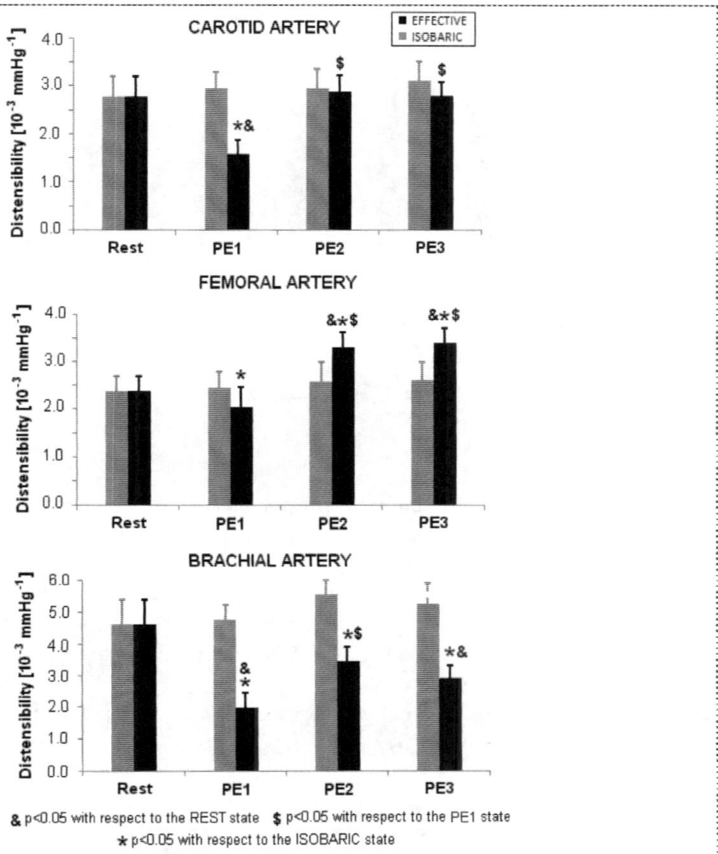

Fig 3: Carotid (top), femoral (middle) and brachial (bottom) effective and passive isobaric distensibility, obtained at rest and after exercise (PE1-PE3).

The PWV basal and post-EST levels are shown in figure 4. Note that EST resulted in an early increase in PWV (higher aorto-iliac arterial stiffness), followed by basal levels recovery.

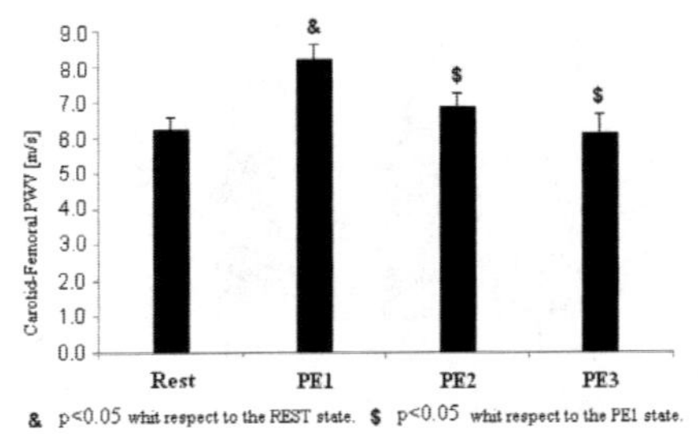

Fig 4: Carotid-femoral pulse wave velocity rest and post-exercise (PE1-PE3) levels.

4. Discussion

In this work and for the first time, the temporal profile of EST-related changes in regional and local biomechanics of elastic, transitional and muscular arteries was characterized. The main results were:

1. The EST resulted in an early (PE1), generalized increase in the arterial stiffness evidenced by both, regional (PWV) and local (distensibility) parameters (Figure 3 and 4).
2. All post-EST recovery stages studied (PE1-PE3) showed quali-quantitative differences in local arterial stiffness (Figure 3).
3. The local arterial biomechanical changes associated with the exercise could not be explained only by blood pressure variations (Figure 3).

Although all the arterial types increased its stiffness immediately after the EST [Figure 2], the changes among the vascular segments were quantitatively different. The increase was 14%, 43% and 57% in the femoral, carotid and brachial arteries, respectively. In addition, considering PE2 and PE3 there were quantitative, but also qualitative differences in the biomechanical response and its temporal profile among the arteries (Figure 3).

While the femoral distensibility increased, reaching values higher than those of rest, the carotid distensibility achieved rapidly rest levels and brachial distensibility did not reach rest levels in the recovery time analyzed.

The dissimilar response could be explained, at least partially, by structural/functional vascular differences and/or by different haemodynamic, EST-related changes in the studied territories that could result in a different capability to respond to exercise (i.e. muscular arteries –femoral- could concentrate the ability to minimize left ventricle afterload and to ensure an adequate lower-limb perfusion during increased conditions of metabolic demands).

As it is known arterial stiffness depends on the arterial pressure, and the exercise is associated with pressure changes (i.e. systolic BP increase and diastolic BP reduction due to, the stroke volume increase and the peripheral dilatation respectively) [2].

To evaluate the pressure-dependence of the stiffness changes associated with EST, the ADE and ADPI were compared. There were differences between both parameters, being the differences highly dependent on the post-EST stage and the vascular segment studied. Looking at these results, the DAE changes could not be ascribed only to BP variations. Instead, pressure-independent, like smooth muscle tone-dependent mechanisms could contribute to explain the distensibility variations found in the post-exercise recovery phase.

5. Conclusion

The EST resulted in regional and local arterial stiffness changes. Local changes were highly dependent on the vascular segment considered. The study of the vascular biomechanics response to EST could add to the CV risk stratification, diagnostic and /or prognostic. About this, further works are necessary to characterize the vascular biomechanical response to EST in different populations and to define its clinical meaning and applicability.

Acknowledgments

We thank Elbio Agote for his technical assistance, Mabel Zabalza for collaborating with the recordings and the students of medicine for participating in the study. We thank ANII for its financial support (FCE 2007-635 and FCE 2007-638)

6. References

[1] Laurent S, Cockcroft J, Van Bortel L, Boutouyrie P, Giannattasio C, Hayoz D, Pannier B, Vlachopoulos C, Wilkinson I, Struijker-Boudier H 2006 Expert consensus document on arterial stiffness: methodological issues and clinical applications. *Eur Heart J*, **27**: 2588-605

[2] Nichols W, O'Rourke M 2005 Properties of the arterial wall: theory Properties of the arterial wall: practice In: Nichols W, O'Rourke M, editors Edward Arnold Publishers, UK *Mc Donald's Blood Flow in Arteries: Theoretical, Experimental and Clinical Principles* 49–65, 67–93

[3] Bia D, Aguirre I, Zócalo Y, Devera L, Cabrera Fischer E, Armentano R 2005 Regional differences in viscosity, elasticity and wall buffering function in systemic arteries: pulse wave analysis of the arterial pressure-diameter relationship *Rev Esp Cardiol*, **58** 167-74

[4] Bia D, Zócalo Y, Armentano R, Laza S, Pérez H, Craiem D, Saldías M, Alvarez I 2009 Non-invasive biomechanical evaluation of implanted human cryopreserved arterial homografts: comparison with preimplanted cryografts and arteries from human donors and recipients *Ann Biomed Eng.* **37**1273-86

[5] Craiem D, Chironi G, Gariepy J, Miranda-Lacet J, Levenson J, Simon A 2007 New monitoring software for larger clinical application of brachial artery flow-mediated vasodilatation measurements. *J Hypertens.* **25** 133-40

[6] Armentano R, Simon A, Levenson J, Chau NP, Megnien JL, Pichel R 1991 Mechanical pressure versus intrinsic effects of hypertension on large arteries in humans *Hypertension* **18**: 657-64.

Journal of Physics: Conference Series **313** (2011) 012014 doi:10.1088/1742-6596/313/1/012014

Learning through projects in the training of biomedical engineers: an application experience

José Antonio Li Gambi and Carmen Peme
Facultad de Ciencias Exactas, Físicas y Naturales. Universidad Nacional de Córdoba. Argentina.

joseligambi@hotmail.com

Abstract. Learning through Projects in the curriculum consists of both the identification and analysis of a problem, and the design of solution, execution and evaluation strategies, with teams of students. The project is conceived as the creation of a set of strategies articulated and developed during a certain amount of time to solve a problem contextualized in situations continually changing, where the constant evaluation provides feedback to make adjustments. In 2009, Learning through Projects was applied on the subject Hospital Facilities and three intervention projects were developed in health centers. This first stage is restricted to the analysis of the aspects that are considered to be basic to the professional training: a) Context knowledge: The future biomedical engineers must be familiarized with the complex health system where they will develop their profession; b) Team work: This is one of the essential skills in the training of students, since Biomedical Engineering connects the knowledge of sciences of life with the knowledge of exact sciences and technology; c) Regulations: The activities related to the profession require the implementation of regulations; therefore, to be aware of and to apply these regulations is a fundamental aspect to be analyzed in this stage; d) Project evaluation: It refers to the elaboration and studying of co-evaluation reports, which helps to find out if Learning through Projects contributes to the training. This new line of investigation has the purpose of discovering if the application of this learning strategy makes changes in the training of students in relation to their future professional career. The findings of this ongoing investigation will allow for the analysis of the possibility of extending its application. Key words: engineering, biomedical, learning, projects, strategies.

1. Introduction

The project is conceived as the creation of a set of strategies articulated among them chronologically, with the purpose of solving a contextualized problem (question, new knowledge, method application) in a dynamic and ever-changing network of situations, in which there is a continuous evaluation that provides feedback and allows us to make the necessary adjustments.

This teaching strategy known as the Project Method has its origin in Kilpatrick's works [1]. Within the curriculum, it constitutes a teaching model focused on the students, who plan, implement and evaluate projects of real world application, beyond the classroom, through activities of interdisciplinary teaching in the long term. The strategy helps to create experiences which turn the learning activity into a process whose purpose is to facilitate knowledge acquisition and the development of theoretical productions, interpretations and contextualized practices. According to the author, the method has four stages: Goal setting, planning, execution and assessment (set the project's goals and objectives; create a plan of tasks and time frames; put it into practice; evaluate the outcomes and draw conclusions). To deal with these situations, the students should apply what they learn in order to solve such situations or suggest improvements in the communities where they take place.

The purpose of the project is not only to learn about something, but also to do something, which implies to acquire knowledge and develop specific skills and a cooperative attitude. In the project, we think about something that does not work well or that could work better, we suggest and evaluate solution alternatives and then we put into practice the one we consider to be the best, taking into account the optimization of costs and time, the maximum exploitation of the resources at stake, and the socio-cultural, environmental and economic impact of the project from the beneficiaries' and potential affected people's point of view. The situations dealt with are changing constantly and there is a continuous evaluation, which provides feedback to make the necessary adjustments [2].

Learning through Projects has evolved from the constructivist theories of learning that belong to Vygotsky, Bruner, Ausubel and other theorists of learning and teaching.

Vygotsky has confirmed through his studies that students better develop their educational processes as they have much more experienced collaborators who allow them to make progress through the zone of proximal development (ZPD) [3], [4]. Learning through Projects is intended to work on and expand the ZPD.

Bruner [5] coined the term scaffolding, which defines the way in which an adult collaborates with a student's learning process. Such process is achieved through the interaction between the subject who learns and the context in which they are placed. The teacher encourages the learner to participate actively. From this comes the emphasis he puts on learning through discoveries, which develops in a context that challenges the students and motivates them to solve problems and transfer what they have learned. The teacher will be then the guide in the discovery. In Learning through Projects, the students find motivating projects in which they play an active role.

García Madruga [6] points out that the students discover the material before they can assimilate it. The students reorganize this material, adjust it to their previous cognitive structure and discover the relationships, laws or concepts that they later assimilate achieving what Ausubel calls significant learning. Learning through Projects has as its purpose to make the students learn deeply and develop specific methodological and social competencies that enable them to act in a competent way in social and work situations.

The teacher's role is to act as a facilitator, a guide, a mediator, a strategist and a negotiator, promoting an atmosphere of participation and cooperation. The teacher interacts with the teams, observes, listens, asks, answers, offers suggestions and stimulates the students' individual and group thinking. They allow the students to behave freely, but they also establish limits, keep expectations and provide guidance for the search of what is essential to know, discuss and shape.

Learning through Projects is a comprehensive methodology that consists of a wide scope of different teaching and learning strategies. If we work with one or more of these strategies, we will create an environment favorable for deepening knowledge and developing competencies and attitudes of all the participants.

In Learning through Projects, the stage in which we evaluate the learning process of the students is very important. In this stage several procedures can be used to determine the achievement of the goals set: a) The initial evaluation allows to diagnose the students' previous knowledge; b) the evaluation based on performance makes it possible to assess the willingness and the learning processes in a specific stage, and to make the necessary adjustments to the educational project; c) the evaluation based on tests or exams helps to estimate the level of the knowledge acquired; d) the auto-evaluation report, oral and written, allows the students to make their own assessment of what they have learned; and e) the evaluation of the project's progress reports helps to assess the improvements the students make and the concepts they learn.

It is necessary that all the projects have one or more public presentations on progress to evaluate the process and the work product in an individual and group manner. This gives the students the opportunity to show what they have learned and increase the validity and authenticity of the evaluation of the prepared project. The progress signs and the medium-term results allow thinking and self-criticism, which can be used to detect problems, analyze difficulties, measure progress and, if necessary, redirect it upon a base of follow-up revisions. However, when the project finishes, we should think about its successes and failures, and the new strategies and actions to adopt in future projects.

2. Background data

The search for background data referred to theoretical frameworks, foundations and a description of the methodology.

In addition, this showed a breaking up in the conceptions about Learning through Projects and the domain in which it is applied. We can find application experiences in school projects (research, technological, social, etc.) in the different levels of the compulsory education (La Cueva [7] and [8], Restrepo [9], MECDV [10] and Castillo Cortés [11]). In the scope of higher education and, more specifically, in professional training courses, the number of works is reduced. The Instituto Tecnológico y de Estudios Superiores de Monterrey ITESM [12] has a significant role as regards this. Application works of Learning through Projects in Biomedical Engineering or in similar courses of study were not found.

This led to an exploratory research with the students that took the subject Hospital Facilities in 2008. Learning through Projects was applied to the topic Medicinal Gases (three units of the program). In the field, it was possible to analyze the development of the method, the students' receptivity, the impact on their learning processes and the difficulties encountered when this project was included in a regular four-month period together with other responsibilities of the students. In this study, aspects to be worked upon were identified, since they would modify the course of study.

3. Context

Strengthening a process started more than 20 years ago with the creation of the Laboratory of Applied Research and Development (L.I.A.D.E.), the Faculty of Exact, Physical and Natural Sciences incorporated Biomedical Engineering to its educational offer. This course of study started to be taught in 2004. The current curriculum (2005) is the result of a joint initiative between the Faculty of Exact, Physical and Natural Sciences and the Faculty of Medical Sciences of the National University of Cordoba (UNC).

Biomedical Engineering is a branch of Engineering, which connects the knowledge of sciences of life with the knowledge of exact sciences and technology, with the purpose of preparing a professional who can work and investigate in the medical and other systems. The graduates will have knowledge of electronics, mechanics, materials and computing; they will also be familiarized with the structure and functioning of hospital systems, which will enable them to suggest solutions to problems related to health prevention and care for the population. The graduates must be able to: a) Design and use sets of instruments of high-complexity technology of biology and medicine, and advise on their use; b) apply the methodology of scientific research to studies within their scope of knowledge; c) participate in the creation of asepsis and safety conditions through the application of measure and radiation equipment; d) collect data for the diagnosis of biological systems through electronic, mechanic, acoustic and optical procedures; e) develop, create and evaluate assistive devices for disabled people; f) study biological systems to develop technological applications; and g) combine multidisciplinary health, research and development equipment.

The curriculum includes four blocks: a) Basic sciences (contents of mathematics, physics, chemistry, biology, computing and design); b) basic technologies (contents of numerical methods, electronics and theory of signals, and specific material of the course of study, such as biomaterials, transducers and sensors); c) applied technologies (contents that constitute the technical knowledge of the course of study and that apply to hospital facilities, biomechanical and biomedical sets of instruments, rehabilitation engineering; the supervised professional practice and the integration project are also included); d) complementary subjects (related to humanistic and philosophical education of the engineer, English and Portuguese as optional subjects, and contents related to management).

To offer the educational service of the subject Hospital Facilities, the coordination department of the School of Biomedical Engineering entrusted the teaching of this subject to the area of Building Facilities and Networks of the Department of Civil Constructions. The subject is developed in the seventh four-month period; that is to say, in the fourth year, with a hourly load of 6 hours per week, a total of 96 hours of classroom activities and a similar quantity of study load and activities not developed in the classroom. In order to register for this subject, students must have passed, or at least have taken, the subjects belonging to the basic technological block, since these will help them to deal with its contents. Given the characteristics of the course of study, the majority of the students are full-time students; a very small percentage of them perform other activities.

4. Definition of problems

Once the background data of the different subjects had been analyzed and the students that took Hospital Facilities had been inquired, it was deduced that Learning through Projects was not used as a strategy during the course of study; therefore, it is not known if its application generates changes in the students' learning, which then could impact on their professional career.

It is important to point out that, in this first stage, the aspects to be studied are limited to the ones considered to be essential for the professional training in this area: a) Knowledge about the context in which the graduates will have to work to be able to develop successfully their profession; b) the specific regulations used; c) team work; and d) the elaboration of reports on the co-evaluation and its subsequent study. Other aspects can also be the subject of new works.

Furthermore, it should be emphasized that this study is restricted to the person to whom the instruction is intended. In further investigations, the subject of study will be the teacher's role, the role of the institution, etc.

According to what was indicated in this study, the following problems are dealt with: 1) Do the group of students who took the subject Hospital Facilities in the first four-month period of 2009, during which Learning through Projects was used, improved their skills in relation to the description they made of the context in which they developed specific projects? 2) Do they make progress as regards their favorable attitude to team work? 3) Do they deepen their knowledge and improve the application of the regulations related to specific projects? 4) Do they master their skills to evaluate projects?

5. Application of Learning through Projects

During the first four-month period of 2009, the 33 students that took Hospital Facilities developed Learning through Projects in three projects which, due to the objectives of the course of study and the scope of the university degree, were treated as intervention projects. The topics developed in the projects were the following: a) fuel gas facilities, b) protection of human beings; fireproof facilities, and c) medicinal gases facilities.

The course was divided in nine work teams, and each team was assigned a health center to visit. For the first project, nine small buildings were chosen: Primary health care facilities, dependent on the Department of Primary Health Care of the Town Hall of the City of Cordoba. For the second project, medium-sized buildings with several medical specialties were chosen, such as clinics that depended upon the same municipal establishment. The last project was developed about big buildings; in this case, the following hospitals were included: Hospital de Niños Santísima Trinidad, Hospital Pediátrico and Hospital Materno Neonatal, dependent upon the Ministry of Health of the Province of Cordoba.

In the projects, the teams had to be aware of the documents that define the provisions and regulations related to the specific facilities of each case. Then, by going to the health center and gathering data about it through records and surveys they considered necessary, they were also able to be in contact with the contexts, the functioning of the health center and the people who work or require medical assistance there.

With the records and data collected, they had to do an analysis of the applicable regulations, and then they had to make an intervention project on the health center, and assess the interventions in the following ways: Urgent, necessary, and desirable. All the information was compiled in one report presented to all students and teachers. The report had to contain, at least, an introduction, which depicted the health center, a description of the current situation in relation to the applicable regulations, and the intervention project.

The oral presentation of each project by the teams also enabled the entire class to be informed about the health center visited and the special characteristics of each case.

For explanatory purposes, two of the developed projects are detailed below:

1.- Team n.° 5, which consists of three students, was assigned Health Center n.° 7, "Pueyrredón". There they had to develop the first intervention project about fuel gas facilities. In their work, they made an introduction to describe the establishment, its dependency, the services provided and the infrastructure it had. Then they made a detailed analysis of the center and verified the compliance with the applicable regulations, in this case, NAG 200. The report finished with the following description: a) Necessary interventions: Correct installation of heating systems in the respective walls and the installation of air intake lines and gas exhausting systems of vent-heaters; b) necessary interventions: Correct installation of pipes, lining of pipes which are outdoors, and installation of ventilation lines and caps for cookers and storage water heaters. At the presentation, the participants expanded the written report and made comments about the special characteristics of the health center in relation to its usefulness for the community.

2.- Team n.° 8, which consisted of three students, was assigned UPAS N° 1, "San Roque". There they had to develop the second project on protection of human beings and fireproof facilities. They carried out the project answering the instructions given and providing a description of the health center, the services provided and the area covered. Then they made an evaluation taking into account the subdivision of the building, fireproofing of its materials, fire exits, fire conditions, and equipment necessary for extinguishing fires. On their final report, they rated the facilities as adequate and optimal, but they pointed out that the fire extinguishers had to be placed in the correct location, since at that opportunity they were kept in a warehouse.

6. Preliminary results

Since this work is limited to the aspects related to the students' learning about the problems pointed out in point IV, the submitted reports were analyzed and the following was evaluated: a) Description of the context through the inclusion of: The place occupied by the health center in the health system, dependency (government level and inside the level Secretary, Department, etc), services provided, community served: Consultations, neighborhoods or areas it serves, socio-economic characteristics of the population; b) team work: The opinion the student had about their work in the project and the opinions of their classmates; the students were asked to complete a chart individually with their own pen name and the one of their classmates (in order to keep the confidentiality of the information), and to rate four analysis categories in a five-point scale: The search of information, role assigned, timely delivery of material prepared, and participation in the report discussion and elaboration; c) as regards the knowledge and application of the specific regulations, certain analysis categories were established in relation to the project. The results for the first and third project were the following: Supply, pipes, devices and facilities, and for the second, presence or absence of the corresponding elements; d) for the projects evaluation, each team was asked to assess the rest of the teams, as regards both the written report and the presentation of the project, previously indicating the evaluation criteria to be used and the reasons of their choices.

The preliminary analysis of the information collected made it possible to identify some progress in the description of the context in each project, an increased participation and commitment to team work, and a very good management of the regulations and their applications.

7. Preliminary conclusions

As regards the application of Learning through Projects, the following conclusions were drawn: a) Important negotiations are required in order to obtain the authorizations that would allow the teams to enter the health centers; b) some problems can arise as the ones occurred with the third project, in which three teams could not visit the hospitals in one occasion due to the social conflicts originated by municipal complaints, and in another opportunity it was considered to be risky that the students enter a hospital due to the influenza A virus; c) the work through projects requires time outside the classroom, and students find it difficult to combine it with the demands for other subjects; and d) students believe Learning through projects is very positive, since it helps them to work in an environment with which they will be in contact during their professional career.

It is relevant to point out the impact this project had on the context of some health centers of the Province of Cordoba. At the moment of selecting the staff to gather data about the medicinal gases facilities, the company in charge of maintenance of medicinal gases facilities in public hospitals established certain selection criteria: the task must be performed by students of this course of study, these students must have taken this subject, they must have applied Learning through Projects, and they must have obtained a very good final grade. Two of the students of this class group were hired for the task.

These conclusions are preliminary and restricted to the aspects studied and to the person to whom the instruction is intended. If we continue with this line of investigation, we will be able to look deeply into other aspects, other roles and the impact of the application of Learning through Projects on the professional career.

Acknowledgements
To the students of the course of study Biomedical Engineering of the Faculty of Exact, Physical and Natural Sciences, National University of Cordoba, who, in the first four-month period of 2008 and 2009, took the subject Hospital Facilities, for their job and cooperation.

To the engineers Carlos Olmos and Walter Gómez, who facilitated the entry of the students to the health centers, and accompanied them and advised them during their work.

To the authorities, medical teams and technicians of the Department of Primary Health Care, Town Hall of the City of Cordoba, for their willingness to cooperate with the students.

References

[1] W. H. Kilpatrick, "The project method", *Teachers college record*, vol. XIX, Nº 4, pp. 319-35, 1918.

[2] C. Peme – Aranega, "Los espacios curriculares integradores en la formación agropecuaria", *Conferencia: 1er. Encuentro. Facultad de Ciencias Agropecuarias,* Universidad Nacional de Rosario, 2007.

[3] L. Vygotsky, "Métodos en la investigación reflexológica y psicológica", 2° *Congreso Panruso de Psiconeurología*, Leningrado, Rusia, 1924.

[4] C. Coll, *Psicología y Currículo*, Laia, Barcelona, 1987.

[5] J. Bruner, "The role of tutoring in problem solving", *The Journal of Child Psychology and Psychiatry, Published on behalf of the Association for Child and Adolescent Mental Health*, V 17, I 2, pp. 89 – 100, 2006.

[6] J. A. García Madruga "Aprendizaje por descubrimiento frente a aprendizaje por recepción: La teoría del aprendizaje verbal significativo", *en Coll, C. Desarrollo psicológico y educación II. Psicología de la educación*, Alianza, Madrid, 1990.

[7] A. La Cueva, "Retos y propuestas para una didáctica contextualizada y crítica", *Educación y Pedagogía*, IX(18), pp.39-82, 1997.

[8] A. La Cueva, "La enseñanza por proyectos ¿mito o reto?", *Revista Iberoamericana de Educación,* N° 16, pp.165-187. 1998.

[9] A. E. Restrepo, "Estrategias para la intervención pedagógica en los procesos de enseñanza de la escritura y la lectura", *Centro de servicios pedagógicos,* Universidad de Antioquia, Medellín, 2000.

[10] Ministerio de Educación, Cultura y Deportes (MECDV), "Metodología del proyecto. Renovemos la escuela Básica, Rural, Indígena y de frontera", *Teoría y praxis*, Venezuela, 2002.

[11] J. Castillo Cortés, "El Aprendizaje por Proyectos", *Red Académica (www.redacademica.edu.co) Publicación de Maestros*, Colombia, 2006.

[12] Instituto Tecnológico y de Estudios Superiores de Monterrey (ITESM), "El método de proyectos como técnica didáctica", Dirección de Investigación y Desarrollo Educativo, Vicerrectoría Académica. http://www.sistema.itesm.mx/va/dide/inf-doc/estrategias/, 1999

AUTHOR INDEX

Institute of Physics
Dirac House, Temple Back
Bristol BS1 6BE UK

CURRAN ASSOCIATES INC
proceedings.com